AF301023

Koichiro Ishibashi
Kenichi Osada

Editors

Low Power and Reliable SRAM Memory Cell and Array Design

With 141 Figures

 Springer

Editors
Prof. Koichiro Ishibashi
The University of Electro-Communications
1-5-1 Chofugaoka, Chofu, Tokyo, 182-8585 Japan
ishibashi@ee.uec.ac.jp

Dr. Kenichi Osada
Hitachi Ltd.
Higashi-koigakubo 1-280, 185-8601 Kokubunji-shi, Tokyo, Japan
kenichi.osada.aj@hitachi.com

Series Editors:
Dr. Kiyoo Itoh
Hitachi Ltd., Central Research Laboratory, 1-280 Higashi-Koigakubo
Kokubunji-shi, Tokyo 185-8601, Japan

Professor Thomas Lee
Stanford University, Department of Electrical Engineering, 420 Via Palou Mall, CIS-205
Stanford, CA 94305-4070, USA

Professor Takayasu Sakurai
Center for Collaborative Research, University of Tokyo, 7-22-1 Roppongi
Minato-ku, Tokyo 106-8558, Japan

Professor Willy M. C. Sansen
Katholieke Universiteit Leuven, ESAT-MICAS, Kasteelpark Arenberg 10
3001 Leuven, Belgium

Professor Doris Schmitt-Landsiedel
Technische Universität München, Lehrstuhl für Technische Elektronik
Theresienstrasse 90, Gebäude N3, 80290 München, Germany

Springer Series in Advanced Microelectronics   ISSN 1437-0387
ISBN 978-3-642-27018-5   ISBN 978-3-642-19568-6 (eBook)
DOI 10.1007/978-3-642-19568-6
Springer Heidelberg Dordrecht London New York

Cover design: eStudio Calamar Steinen

Printed on acid-free paper

Springer is part of Springer Science+Business Media (www.springer.com)

Preface

As LSI industry has been growing, there appear such kinds of CMOS LSI as Microprocessor, MCU, gate array, ASIC, FPGA, and SOC. There is no CMOS LSI that does not use the SRAM memory cell arrays. The best reason is that the SRAM can be fabricated by the same process as logic process, and it does not need extra cost to fabricate. Also, SRAM cell array operates fast and consumes low power in LSI. Despite the SRAM cell size is larger than the other RAM cell such as DRAM cell and Flash, SRAM cell continue to be used in CMOS LSI, thanks to the natures of the cell.

Before 90 nm technology, we can design SRAM cell for CMOS LSI without paying attention to electrical stability, so that we could get operable SRAM bit cell only when we connect the six transistors of the cell. However, after 90 nm technology, we must design SRAM bit cell more carefully, because variability and leakage of transistors in SRAM cell have become large. Also, we must pay attention to such reliability issues as soft errors, and NBTI at low supply voltages.

This book is focusing on the design of CMOS memory cell and memory cell array, taking low voltage operation and reliability into consideration. The authors are specialists who have engaged in these issues for tens of years in Hitachi Ltd, and Renesas Technology Corporation that is currently Renesas Electronics Corporation. I believe this book can help readers understand fundamentals of CMOS SRAM memory cell and cell array design, design methods of memory cell, and cell array taking variability of transistors into considerations, thereby obtaining low power and reliable SRAM arrays. This book also introduces new memory cell design techniques those we can apply to future LSI technologies such as SOI devices.

Acknowledgments

The editors and authors express special thanks to Dr. Kiyoo Itoh of Hitachi Ltd. for encouragement in editing this book. We also appreciate Dr. Toshiaki Masuhara, Dr. Osamu Minato, Mr. Toshio Sasaki, and Dr. Toshifumi Shinohara for leading the authors in various SRAM and SOC development projects. We would like to appreciate many colleagues, who have been working on the various projects with us.

Tokyo, April 2011   *Koichiro Ishibashi*
                                                      Kenichi Osada

Contents

1 Introduction 1
Koichiro Ishibashi
1.1 History and Trend of SRAM Memory Cell 1
1.2 Memory Cell Design Techniques and Array Design Techniques 3
References 4

2 Fundamentals of SRAM Memory Cell 5
Kenichi Osada
2.1 SRAM Cell 5
2.2 Basic Operation of SRAM Cell 5
2.3 Electrical Stability at Read Operation:
Static Noise Margin and β Ratio 9
Reference 10

3 Electrical Stability (Read and Write Operations) 11
Masanao Yamaoka and Yasumasa Tsukamoto
3.1 Fundamentals of Electrical Stability
on Read and Write Operations 11
3.2 V_{th} Window Curve 16
3.3 Sensitivity Analysis 19
References 24

4 Low Power Memory Cell Design Technique 25
Kenichi Osada and Masanao Yamaoka
4.1 Fundamentals of Leakage of SRAM Array 25
4.1.1 Leakage Currents in an SRAM of Conventional Design 26
4.1.2 Gate-Tunnel Leakage and GIDL Currents 26
4.2 Source Line Voltage Control Technique 29
4.2.1 EFR Scheme for Low Power SRAM 29
4.2.2 Chip Architecture 29
4.2.3 Results 31

4.2.4   Source Line Voltage Control Technique
for SRAM Embedded in the Application Processor 32
4.3   LS-Cell Design for Low-Voltage Operation 36
4.3.1   Lithographically Symmetrical Memory Cell 37
References 40

5 Low-Power Array Design Techniques 43
Koji Nii, Masanao Yamaoka, and Kenichi Osada
5.1   Dummy Cell Design 45
5.1.1   Problem with Wide-Voltage Operation 45
5.1.2   Block Diagram and Operation of Voltage-Adapted
Timing-Generation Scheme 46
5.1.3   Timing Diagram and Effect of Voltage-Adapted
Timing-Generation Scheme 48
5.1.4   Predecoder and Word-Driver Circuits 50
5.1.5   Results.. 51
5.2   Array Boost Technique 54
5.3   Read and Write Stability Assisting Circuits 59
5.3.1   Concept of Improving Read Stability 59
5.3.2   Variation Tolerant Read Assist Circuits 62
5.3.3   Variation Tolerant Write Assist Circuits 67
5.3.4   Simulation Result 71
5.3.5   Fabrications and Evaluations in 45-nm Technology 72
5.4   Dual-Port Array Design Techniques 74
5.4.1   Access Conflict Issue of Dual-Port SRAM 74
5.4.2   Circumventing Access Scheme of Simultaneous
Common Row Activation 76
5.4.3   8T Dual-Port Cell Design 80
5.4.4   Simulated Butterfly Curves for SNM 81
5.4.5   Cell Stability Analysis... 83
5.4.6   Standby Leakage... 84
5.4.7   Design and Fabrication of Test Chip........................... 84
5.4.8   Measurement Result 86
References 87

**6 Reliable Memory Cell Design for Environmental
Radiation-Induced Failures in SRAM** 89
Eishi Ibe and Kenichi Osada
6.1   Fundamentals of SER in SRAM Cell 90
6.2   SER Caused by Alpha Particle 94
6.3   SER Caused by Neutrons and Its Quantification 97
6.3.1   Basic Knowledge of Terrestrial Neutrons 97
6.3.2   Overall System to Quantify SER–SECIS....................... 99
6.3.3   Simulation Techniques to Quantify Neutron SER 99
6.3.4   Predictions of Scaling Effects from CORIMS 102

6.4 Evolution of MCU Problems and Clarification of the Mechanism ... 105
    6.4.1 MCU Characterization by Accelerator-Based
          Experiments 105
    6.4.2 Simplified 3D Device Simulation Mixed
          with Circuit Simulation 108
    6.4.3 Full 3D Device Simulation with Four-Partial-Cell
          Model and Multi-Coupled Bipolar Interaction (MCBI) 112
6.5 Countermeasures for Reliable Memory Design 115
    6.5.1 ECC Error Correction and Interleave Technique
          for MCU 115
    6.5.2 ECC Architecture 117
    6.5.3 Results 118
References 119

7 Future Technologies 125
  Koji Nii and Masanao Yamaoka
  7.1 7T, 8T, 10T SRAM Cell 125
  7.2 Thin-Box FD-SOI SRAM 128
  7.3 SRAM Cells for FINFET 135
  References 137

Index 139

Contributors

Eishi Ibe, Production Engineering Research Laboratory, Hitachi, Ltd., 292 Yoshida, Totsuka, Yokohama, Kanagawa 244-0817, Japan, hidefumi.ibe.hf@hitachi.com

Koichiro Ishibashi, The University of Electro-Communications, 1-5-1 Chofu-gaoka, Chofu, Tokyo, 182-8585 Japan, ishibashi@ee.uec.ac.jp

Koji Nii, Renesas Electronics Corporation, 5-20-1, Josuihon-cho, Kodaira, Tokyo 187-8588, Japan, koji.nii.uj@renesas.com

Kenichi Osada, Measurement Systems Research Department, Central Research Laboratory, Hitachi, Ltd., 1-280, Higashi-koigakubo, Kokubunji-shi, Tokyo 180-8601, Japan, kenichi.osada.aj@hitachi.com

Yasumasa Tsukamoto, Renesas Electronics Corporation, 5-20-1, Josuihon-cho, Kodaira, Tokyo 187-8588, Japan, yasumasa.tsukamoto.gx@renesas.com

Masanao Yamaoka, 1101 Kitchawan Road, Yorktown Heights, NY 10598, USA, masanao.yamaoka@hal.hitachi.com

Chapter 1
Introduction

Koichiro Ishibashi

1.1 History and Trend of SRAM Memory Cell

Static random access memory (SRAM) has been widely used as the representative memory for logic LSIs. This is because SRAM array operates fast as logic circuits operate, and consumes a little power at standby mode. Another advantage of SRAM cell is that it is fabricated by same process as logic, so that it does not need extra process cost. These features of SRAM cannot be attained by the other memories such as DRAM and Flash memories. SRAM memory cell array normally occupies around 40% of logic LSI nowadays, so that the nature of logic LSI such as operating speed, power, supply voltage, and chip size is limited by the characteristics of SRAM memory array. Therefore, the good design of SRAM cell and SRAM cell array is inevitable to obtain high performance, low power, low cost, and reliable logic LSI.

Various kinds of SRAM memory cell has been historically proposed, developed, and used. Representative memory cell circuits are shown in Fig. 1.1.

High-R cell was first proposed as low power 4 K SRAM [1.1]. In the High-R cell, high-resistivity poly-silicon layer is used as load of inverter in the SRAM cell. The High-R cell does not need bulk PMOS, so that the memory cell size was smaller than 6-Tr. SRAM. As the resistivity of the poly-silicon layer is around 10^{12}, the standby current of the memory cell was dramatically decreased to 10^{-12} per cell. The high-R cell was widely used for high density and low power SRAM memory LSI from 4 K to 4 M bit [1.2, 1.3]. The disadvantage of the high-R cell is low voltage operation. At low supply voltages less than 1.5 V, the cell node voltage should be charged to supply voltage level during write operation. Since the resistivity of the load poly-silicon is high, the time required to charge up the high node to supply voltage level is quit large, high-R cell cannot operate at supply voltages less than 1.5 V.

K. Ishibashi
The University of Electro-Communications, 1-5-1 Chofugaoka, Chofu, Tokyo, 182-8585 Japan
e-mail: ishibashi@ee.uec.ac.jp

K. Ishibashi and K. Osada (eds.), *Low Power and Reliable SRAM Memory Cell and Array Design*, Springer Series in Advanced Microelectronics 31,
DOI 10.1007/978-3-642-19568-6_1, © Springer-Verlag Berlin Heidelberg 2011

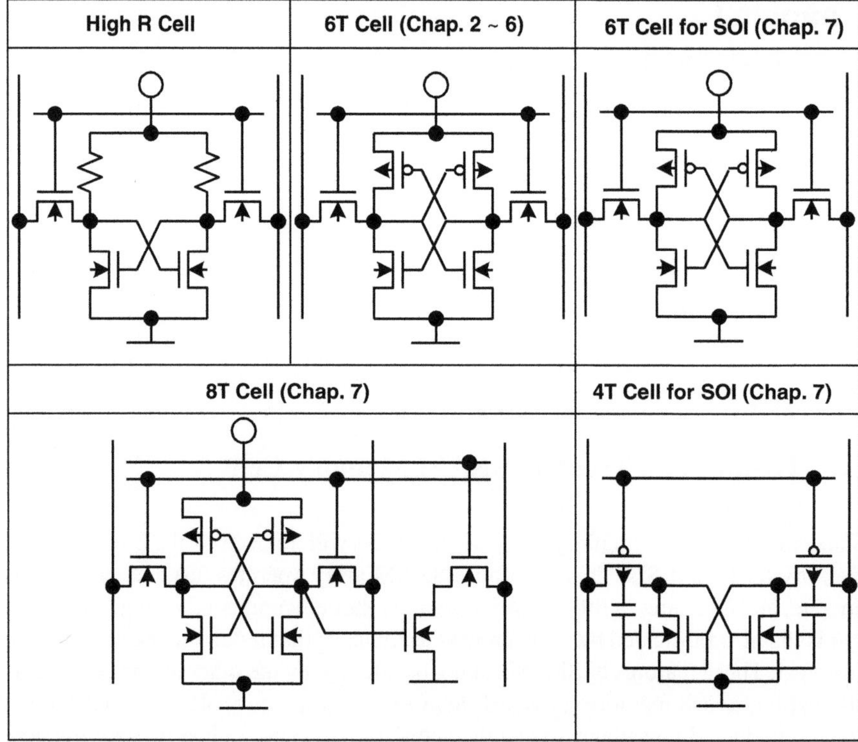

Fig. 1.1 Representative SRAM memory cell circuits

Six-transistor cell (6T cell), which is sometimes called as full CMOS cell, has been widely used as memories for logic LSIs instead of high-R cell. Most parts of this book deals with this type of memory cell. Although the 6T cell uses PMOS transistors and cell area becomes larger than high-R cell, the cell does not need extra process to logic process. Hence, it has been widely used for memories for logic LSIs even during high-R cell was popular for standalone SRAM. In addition, the PMOS transistors in the cell pull up the cell nodes voltages fast, so that 6T cell operates at lower supply voltages than high-R cell. Therefore, recent supply voltage reduction at advanced technologies has made the 6T cell inevitable for logic LSI. Figure 1.2 shows the size of 6T cell in recent VLSI symposia and International Electron Devices Meeting (IEDM). The cell size has been decreasing by half every 2 years, and it corresponds to density enhancement by Moore's law. Therefore, 6T cell is main-stream memory cell for past, nowadays, and future logic LSIs. Even though structure of transistor will be changed to SOI or FINFET, the 6T cell could be used as the main memory cell for logic LSI.

For further needs to extremely low supply voltage and fast operation, eight-transistor cell (8T cell) has been proposed. Moreover, such special four-transistor

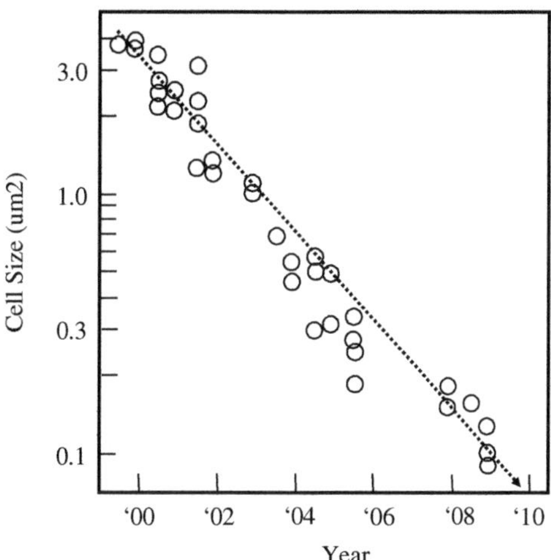

Fig. 1.2 6T cell size trend in VLSI and IEDM

cell (4T cell) has been proposed using FINFET transistor structure. These kinds of new SRAM memories are treated in Chap. 7 as future technologies of SRAM cell.

1.2 Memory Cell Design Techniques and Array Design Techniques

There are a lot of issues to obtain low power, reliable, and small cell size 6T memory cell. Since the 6T SRAM cell size is scaled by Moore's law, the feature size of transistors in 6T cell is also reduced by Moore's law. Supply voltage of 6T cell is also reduced as the feature size is reduced. Variation in the transistors' threshold voltage has increased and leakage of transistors has also increased by the scaling. Supply voltage of memory cell array has been reduced by the scaling. Recent low power circuit techniques such as Dynamic Voltage Frequency Scaling (DVFS) alsoneed further low voltage operation of memory cell arrays.

The 6T SRAM cell must be designed so that it must be electrically stable at the low supply voltages despite the large variation of transistors. The memory cell size must be as small as possible to obtain small chip size LSIs. The leakage of 6T SRAM cell array must be small despite large leakage in transistors in the cell. In addition, immunity to soft errors due to alpha particles or neutrons must be minimized to obtain reliable LSIs.

This book is focusing on design techniques of SRAM memory cell and array, and covers issues on variability, low power and low voltage operation, reliability, and future technologies.

This book first explains electrical stability issue as fundamentals of electrical operation of 6T cells in Chap. 2. Precise analysis techniques of electrical stability, V_{th} window analysis, and sensitivity analysis will be introduced in Chap. 3. Using the analysis techniques, suitable V_{th} for PMOS and NMOS transistors in 6T cell are determined, so that electrically stable 6T cell at low supply voltages under large variability circumstances is obtained.

The SRAM cell array must operate at low voltage operation to reduce operating power. It must retain data with low leakage at standby mode. Many low power techniques for obtaining low power SRAM have been proposed. This book covers important low power memory cell array design techniques as well as memory cell design techniques.

Two important memory cell design techniques will be introduced in Chap. 4. Lithographically Symmetric Cell (LS cell) is symmetric memory cell layout, so that balance of characteristics in the paired MOS transistors in 6T cell, and good electrical stability can be obtained with advanced super-resolution photolithography. Source voltage control technique, which can reduce not only subthreshold leakage but also gate-induced drain leakage (GIDL), will also be shown to reduce data retention current in standby mode.

SRAM cell array design plays also an important role in reducing power consumption. Dummy cell design technique to adjust activation timing of sense amplifier will be shown in Chap. 5, so that stable SRAM operation is achieved with PVT variation circumstances. Assisting circuits at read operation as well as write operation will be proposed to attain low voltage operation of memory cell arrays. Array boost technique is also shown to obtain the lowest operation voltage of 0.3 V.

Reliability issue is another inevitable issue for SRAM memory cell and cell array design. Among various reliability issues, soft errors caused by alpha particles and neutron particles are serious issue. This book first shows the phenomena of the soft error on SRAM memory cell array and explains mechanisms of the soft errors in Chap. 6. Then memory cell array designtechniques drastically reduce the soft errors.

Chapter 7 is the final chapter of this book. This chapter introduces future design techniques of SRAM memory cell and array. Such SRAM cell with such larger number of transistors as 8T cell will be shown to obtain SRAM array with lower supply voltages. Then SRAM memory cell using SOI and FINFET technology will be discussed for future design techniques.

References

1.1. T. Masuhara et al., A high speed, low-power Hi-CMOS 4 K static RAM, in *IEEE International Solid-State Circuits Conference, Digest* 1978, pp. 110–111
1.2. O. Minato et al., A 42 ns 1 Mb CMOS SRAM, in *IEEE International Solid-State Circuits Conference, Digest* 1987, pp. 260–261
1.3. K. Sasaki et al., A 23 nm 4 Mb CMOS SRAM, in *IEEE International Solid-State Circuits Conference, Digest* 1990, pp. 130–131

Chapter 2
Fundamentals of SRAM Memory Cell

Kenichi Osada

Abstract This chapter introduces fundamentals of SRAM memory cell. The basic SRAM cell design and the operation are also described in this chapter. In Sect.2.1, the most common SRAM cell, the full CMOS 6-T memory cell, is explained. In Sect.2.2, read and write basic operations are introduced. In Sect.2.3, the basic of electrical stability at read operation (static noise margin, SNM) is described.

2.1 SRAM Cell

The SRAM cell is constituted of a flip-flop. On the storage nodes of the flip-flop, logical data "0" or "1" is stored. The most commonly used SRAM cell is the full CMOS 6-transistor (6-T) memory cell as shown in Fig. 2.1. The SRAM cell consists of two inverters [load MOSFET(LD0)-driver MOSFET(DR0), LD1-DR1], and two access MOSFETs (AC0, AC1) that are connected to a pair of bit lines (BT, BB). The two access MOSFETs are also connected to a word line (WL). To form a flip-flop, the input and the output of one inverter are connected to the output and the input of the other inverter, respectively. As far as system-on-chip (SoC) design is considered, the full CMOS 6-T memory cell is mostly used since it is fabricated by the same process as CMOS logic uses.

2.2 Basic Operation of SRAM Cell

Figure 2.2 shows a basic full CMOS cell array structure. The memory array consists of n by m bits of memory cells. A word decoder selects one WL based on addresses (A). Column control circuits consisting of precharge circuit, column

K. Osada
Measurement Systems Research Department, Central Research Laboratory, Hitachi Ltd.,
1-280, Higashi-koigakubo, Kokubunji-shi, Tokyo 180-8601, Japan
e-mail: kenichi.osada.aj@hitachi.com

K. Ishibashi and K. Osada (eds.), *Low Power and Reliable SRAM Memory Cell and Array Design*, Springer Series in Advanced Microelectronics 31,
DOI 10.1007/978-3-642-19568-6_2, © Springer-Verlag Berlin Heidelberg 2011

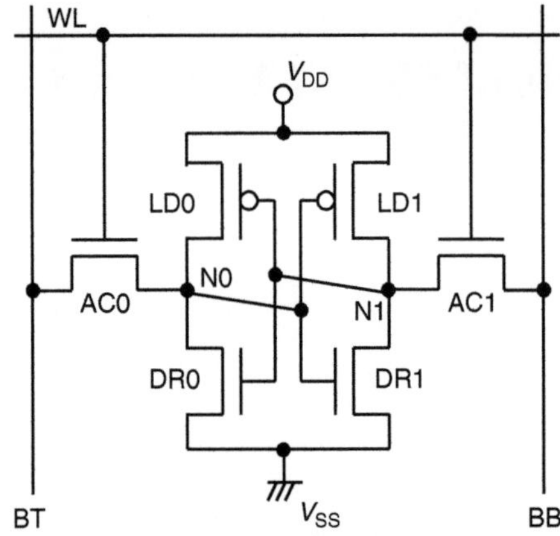

Fig. 2.1 SRAM cell with the full CMOS 6-transistor (6-T)

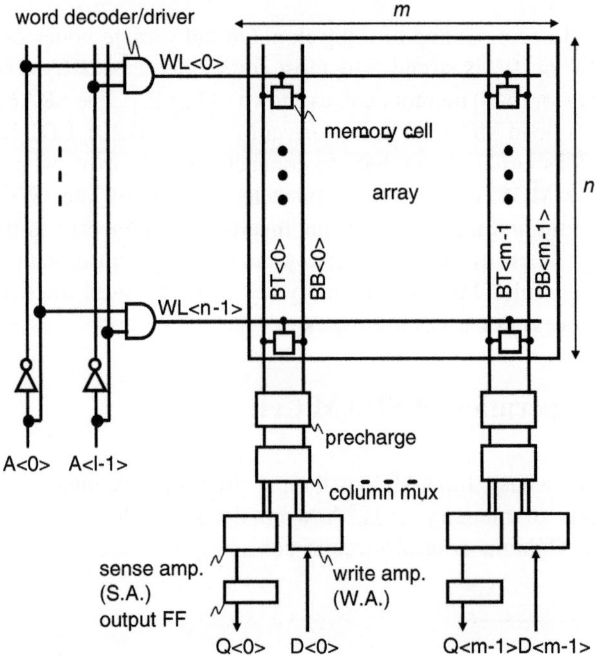

Fig. 2.2 Basic full CMOS cell array structure

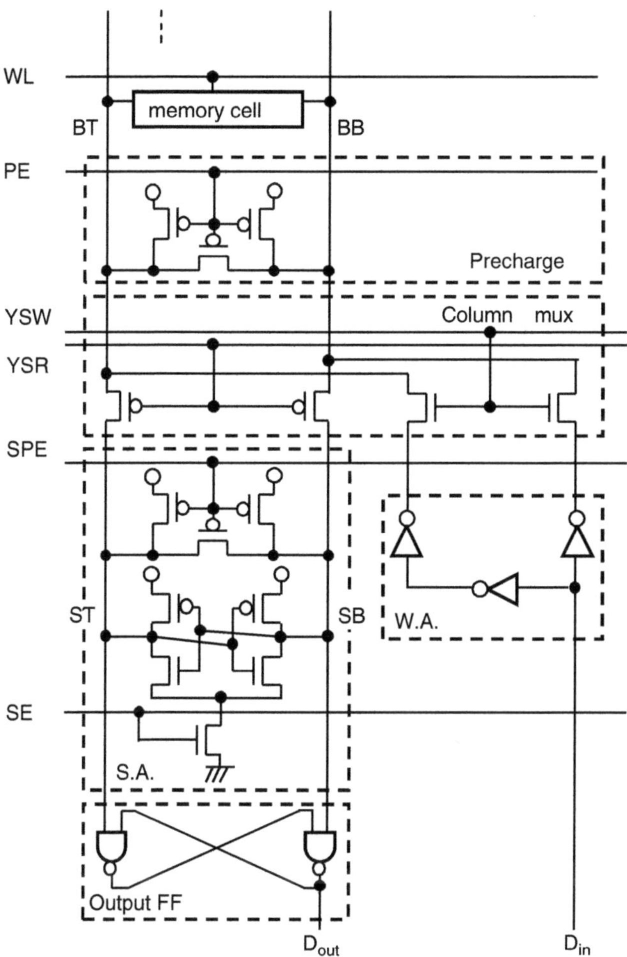

Fig. 2.3 Column control circuits consisting of precharge circuit, column multiplexer (mux), sense amplifier (SA), output FF, and write amplifier (WA)

multiplexer (mux), sense amplifier (SA), output FF, and write amplifier (WA) are shown in Fig. 2.3. The basic voltage waveform in read and write operation is shown in Fig. 2.4. The read operation starts with activation of word line selection. Before the activation, the precharge circuit equalizes and raised the bit-line voltages to V_{DD} level. Each cell generates a small signal voltage, ΔV_S, on one of the bit lines, depending on the stored cell data. If stored data is "1," where cell node N0 is at a low voltage and cell node N1 is at a high voltage, the BT voltage is decreased by ΔV_S. The other bit line (BB) remains at the equalized voltage because access transistor (DR1) is not turned on. The differential signals are transferred to a pair of sense-amplifier-input signals (ST/SB) by turning on the column switch selected

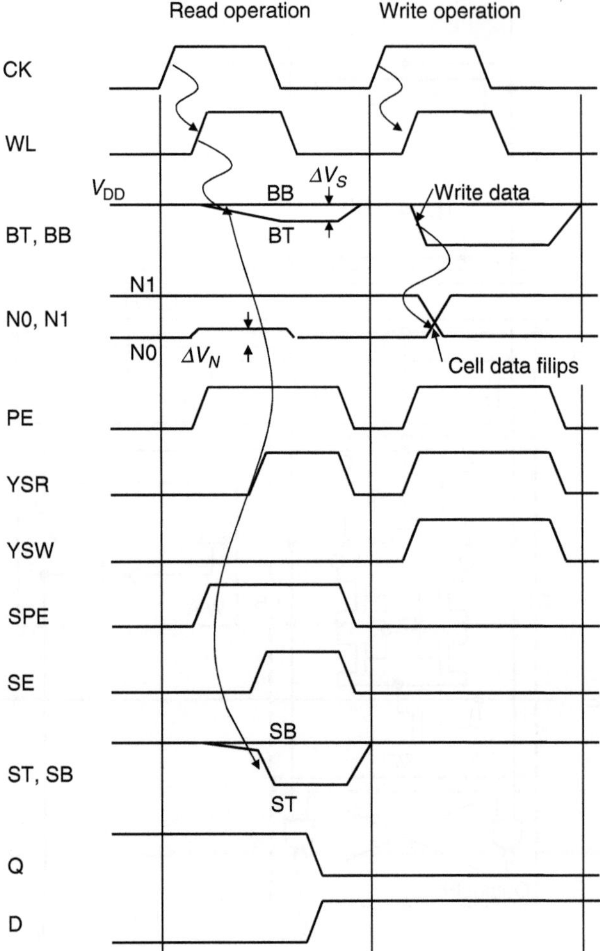

Fig. 2.4 Basic voltage waveform in read and write operation

by a column-select line for read (YSR) in the column mux. The signal voltage is amplified by the sense amplifier. The amplified data are latched by output FF, so as to be outputted as the data output (D_{out}) from the array. When data "0" is stored, the same operation is carried out, except that the BB voltage is decreased.

The write operation starts with activation of a selected word line. Before the activation, the bit lines are equalized to around V_{DD}, as shown in Fig. 2.4. Write data voltage, whose logical level corresponds to the write data on the data input (D_{in}), is applied to the bit lines by the write amplifier (WA). The column switch is turned on by a column-select signal for write (YSW). The write-data voltages are transferred to the bit lines. Then the cell-node voltages (N1, N2) flip. The write operation is completed by turning off the word line.

2.3 Electrical Stability at Read Operation: Static Noise Margin and β Ratio

At normal read operation, when a word line is activated in the case that cell node N0 is at a low voltage and cell node N1 is at a high voltage, the read current flows from BT to N0. Then, the N0 voltage rises by ΔV_N (Fig. 2.4). At abnormal operation, ΔV_N is larger than the DR1 threshold voltage (V_{th}), DR1 turns on, and the N1 voltage decreases. As a result, the N0 voltage increases and the cell data flip. Therefore, we need to design SRAM cell so as to avoid abnormal operation. One important index to evaluate an electrical stability is static noise margin (SNM) [2.1]. This index represents the margin voltage in which data upset occurs during read operation. To estimate the SNM, the word line and a pair of bit lines are connected to V_{DD} shown in Fig. 2.5. The SNM is simulated as follows:

1. While raising the N0 voltage from 0 V to V_{DD}, measure the N1 voltage and N1–N0 relations are plotted.
2. While raising the N1 voltage from 0 V to V_{DD}, measure the N0 voltage and N0–N1 relations are plotted.
3. Draw squares as large as possible between the line A and the line B (Squares C and D). A length of diagonal line of the smaller square is designed as the SNM.

These A and B lines are called as a "butterfly curb." In Fig. 2.6, the butterfly curbs are shown at $V_{DD} = 0.6$ V and 1.1 V. An SRAM cell provides a smaller noise margin at lower voltage. The asymmetry of butterfly curbs increases as the threshold voltage (V_{th}) variation in a cell increases. As a result, the SNM decreases. Lower V_{th} of driver MOSFET decreases the SNM. The ratio of the gate width of driver MOS to that of an access MOS is called β ratio. Larger β ratio increases the SNM. This is because the N0 voltage rise by ΔV_N can be restrained by increasing the driver MOS current and decreasing the access MOS current. Generally, in the SRAM cell, the β ratio is over 1.5.

Because the full CMOS 6-T memory cell statically retains data, it does not need special treatment like refresh. At read operation, fast access time is achieved

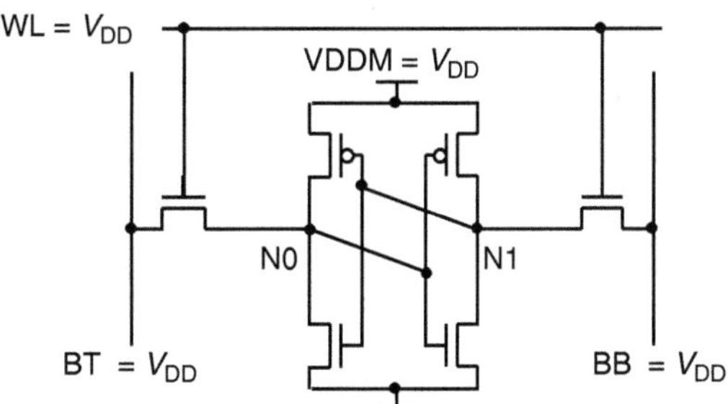

Fig. 2.5 Connections of the word line and a pair of bit lines to estimate the SNM

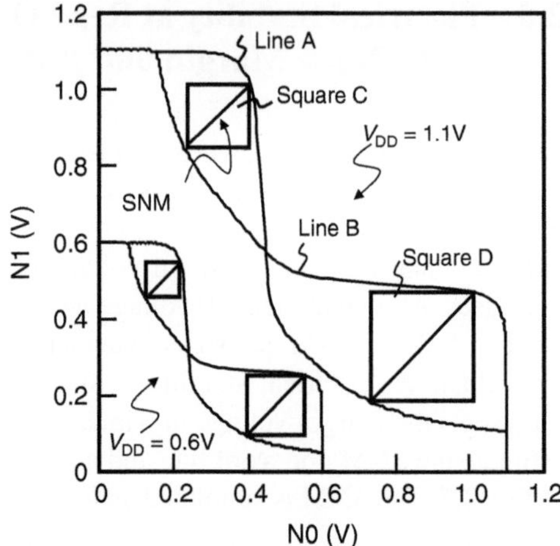

Fig. 2.6 Butterfly curbs are shown at VDD = 0.6 V and 1.1 V

because complementary bit line signals make it possible to use differential amplifier. Therefore, SRAM is used as cache memory. SRAM is also used in mobile phones, e.g., which are required for low retention power. SRAM cell is needed to reduce the power consumption in standby mode during retaining data reliably.

Reference

2.1. E. Seevinck, F.J. List, J. Lohstroh, Static-noise margin analysis of MOS SRAM cells. IEEE J. Solid-State Circuits **SC-22**(5), 748–754 (1987)

Chapter 3
Electrical Stability (Read and Write Operations)

Masanao Yamaoka and Yasumasa Tsukamoto

Abstract In SRAM, read and write are fundamental operations. To ensure the correct operations, the stability analysis is indispensable. In this chapter, electrical stability analysis is explained. In Sect. 3.1, the SRAM operations, read and write, are explained. In this section, the read stability, static-noise margin SNM, and the write stability are described. In SRAM design, V_{th} variation of transistors has critical influence on SRAM operation. In Sect. 3.2, the V_{th} variation of MOSFETs and its effect to SRAM are described. The V_{th} variations can be divided into local and global components. In this section, the effect of V_{th} variation is made visible using V_{th} window curve analysis. In Sect. 3.3, by means of the conventional SNM and write margin analysis on the SRAM cell characteristics, expanded mathematical analysis to obtain the V_{th} curve is described. The analysis is instructive to see stable V_{th} conditions visually to achieve high-yield SRAM. Furthermore, the proposed analysis referred to as the worst-vector method allows to derive the minimum operation voltage of the SRAM Macros (V_{ddmin}).

3.1 Fundamentals of Electrical Stability on Read and Write Operations

A circuit diagram of a CMOS 6-transistor SRAM memory cell is shown in Fig. 3.1. The fundamental waveform of SRAM operation is indicated in Fig. 3.2. When data "0" is written to the cell (W0 in Fig. 3.2), the bit lines are driven to "H" and "L"

M. Yamaoka (✉)
Hitachi America Ltd., Research and Development Division, 1101 Kitchawan Road, Yorktown Heights, NY 10598, USA
e-mail: masanao.yamaoka@hal.hitachi.com

Y. Tsukamoto
Renesas Electronics Corporation, 5-20-1, Josuihon-cho, Kodaira, Tokyo 187-8588, Japan
e-mail: yasumasa.tsukamoto.gx@renesas.com

K. Ishibashi and K. Osada (eds.), *Low Power and Reliable SRAM Memory Cell and Array Design*, Springer Series in Advanced Microelectronics 31, DOI 10.1007/978-3-642-19568-6_3, © Springer-Verlag Berlin Heidelberg 2011

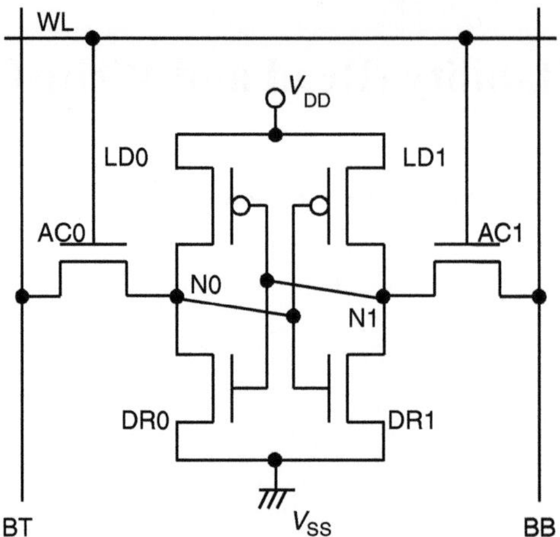

Fig. 3.1 SRAM memory cell

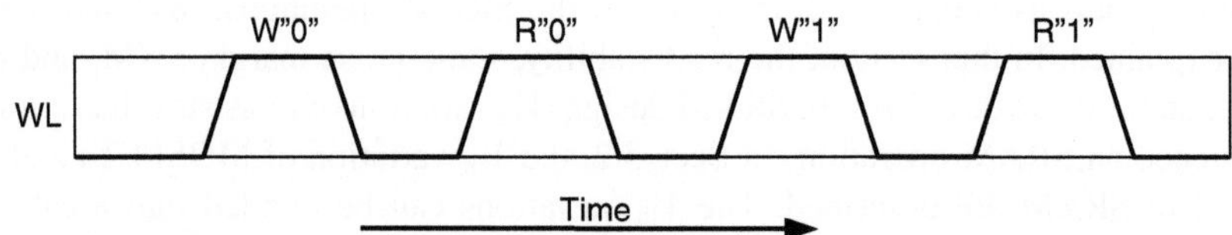

Normal Operation

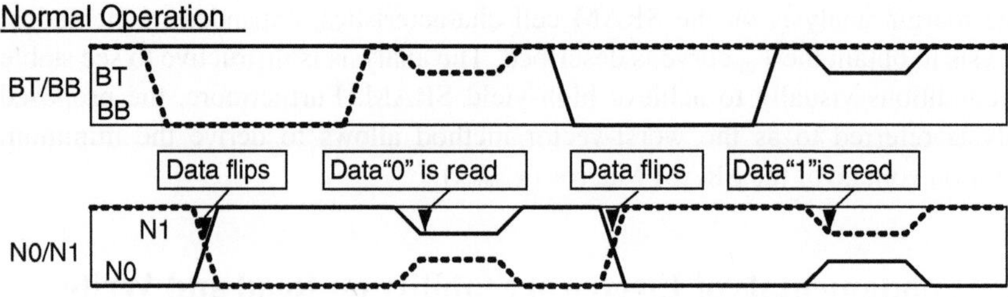

Read stability failure case

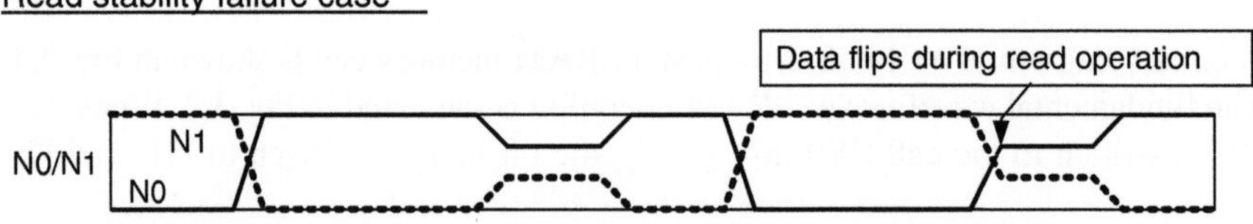

Write failure case

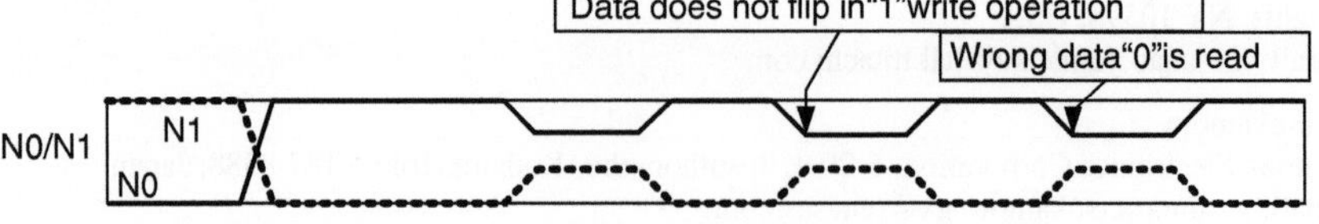

Fig. 3.2 Fundamental waveform of SRAM operation

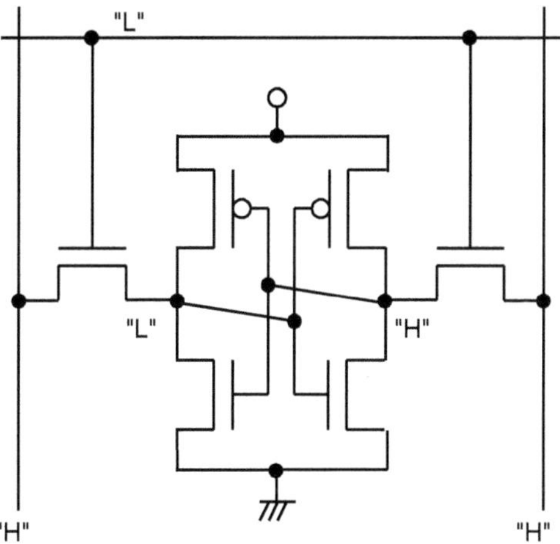

Fig. 3.3 SRAM cell condition during data retention

according to the input data and the word line (WL) is activated. The input data is written to the data storage node, and the voltages of the data storage nodes are flipped. When data "0" is read from the cell (R0 in Fig. 3.2), both bit lines are precharged to "H," and then the WL is activated. The retained data is appeared to bit lines and voltage difference on the bit lines is generated. The voltage difference is amplified at sense amplifier and the retained data is read out. When data "1" is written to the cell and write failure is occurred (W1 write failure case in Fig. 3.2), the voltages of the storage nodes (N0/N1) are not flipped, and the correct data is not written into the cell. At the next read cycle, incorrect data "0" is read out. When data "1" is read out from the cell and read failure is occurred (R1 read failure case in Fig. 3.2), the voltage of data storage nodes (N0/N1) is flipped and the retained data is destroyed. Therefore, the read failure is occurred.

Figure 3.3 shows a condition where SRAM memory cell retains a data. One memory cell has two data storage nodes, N0 and N1, and each node becomes the opposite state, "H" and "L." A word line is "L" state.

In Fig. 3.4, a read operation of a memory cell, the right storage node, N1, is "H" state, and the left storage node, N0, is "L", is indicated. Before read operation, the bit lines are precharged to "H" state. In read operation, the word line becomes "H" state, and the access MOSs are turned on. A read cell current I_{read} flows through the access MOS and "L"-state node and discharges the precharged bit line indicated as an arrow in Fig. 3.4.

Figure 3.5a shows the voltages of each node in SRAM cell during normal read operation. When the read operation is normal, the storage nodes' voltages are not flipped. In Fig. 3.5b, the butterfly curves of normal-operation memory cell, which is used to estimate the static noise margin (SNM), are indicated. There are spaces between two lines and the SNM is sufficient for SRAM read operation. During read operation, the voltage of "L"-state storage node is raised due to the resistive division between access MOS and driver MOS. If the raised voltage of "L"-state storage node becomes higher than the logical threshold voltage of the inverter, composed of driver

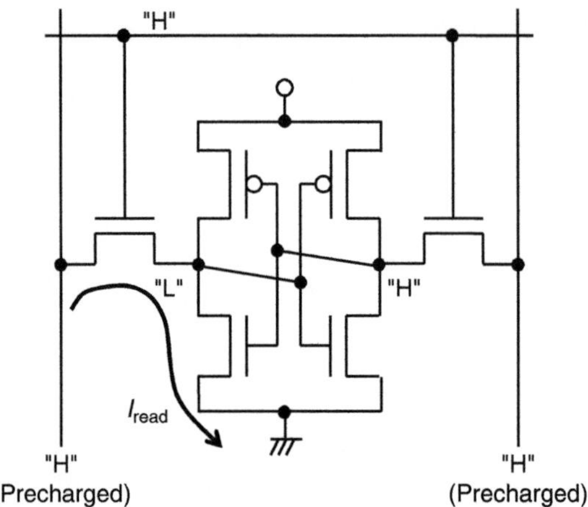

Fig. 3.4 Read operation of SRAM cell

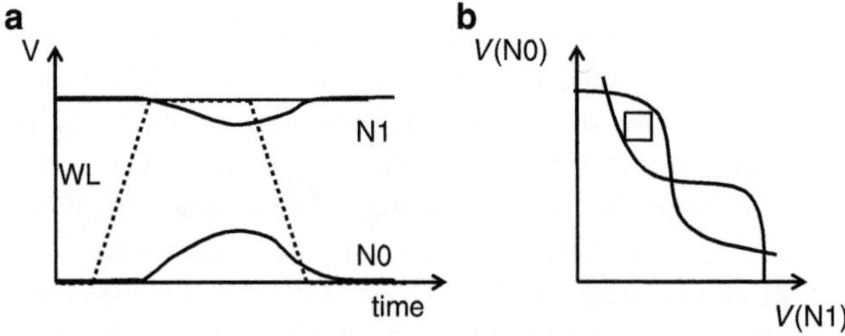

Fig. 3.5 Voltage waveforms and butterfly curves during normal read operation. (**a**) Voltage waveforms of SRAM cell nodes during normal read operation. (**b**) Butterfly curves to estimate SNM

MOS, DR1, and load MOS, LD1, the retained data are flipped and destroyed, as indicated in Fig. 3.6a. This situation indicates a failure read operation. In Fig. 3.6b, the butterfly curves of failure read operation memory cell are indicated. There is only one space between two lines, and the SNM is insufficient for normal SRAM read operation.

Figure 3.7 shows write operation. Before beginning of the write operation, the left storage node, N0, is "H" state, and the right storage node, N1, is "L" state. When the write data come from bit lines, the electrical charge in the "H" potential node is discharged through the access MOS by write cell current I_{write}, indicated as a solid arrow in the figure. Write operation is completed by discharging the "H" potential node to "L" potential and the nodes' voltages are flipped as shown in Fig. 3.8.

During the write operation, the load MOS supplies electrical charge to the "H" potential node by pull-up current I_{pullup}, indicated as a dashed arrow in Fig. 3.7. Therefore, if the charging current of the load MOS is larger than the discharging current of the access MOS, the "H" potential cannot be discharged, and the nodes' voltages are not flipped and write operation fails as shown in Fig. 3.9.

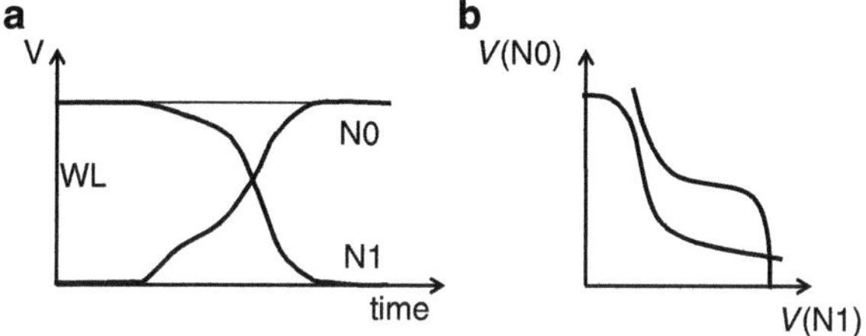

Fig. 3.6 Voltage waveforms and butterfly curves during failure read operation. (**a**) Voltage waveforms of SRAM cell nodes during failure read operation. (**b**) Butterfly curves to estimate SNM

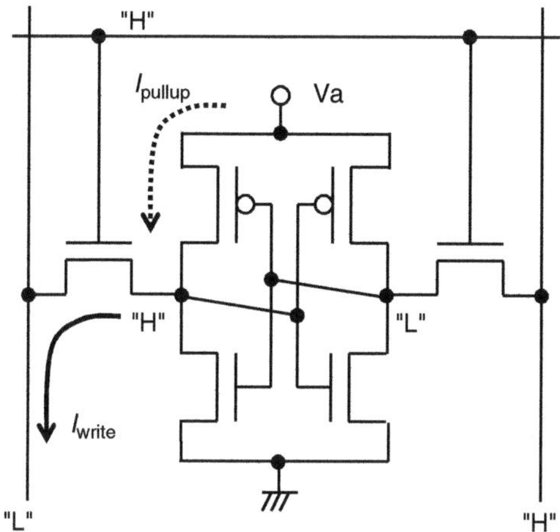

Fig. 3.7 Write operation of SRAM cell

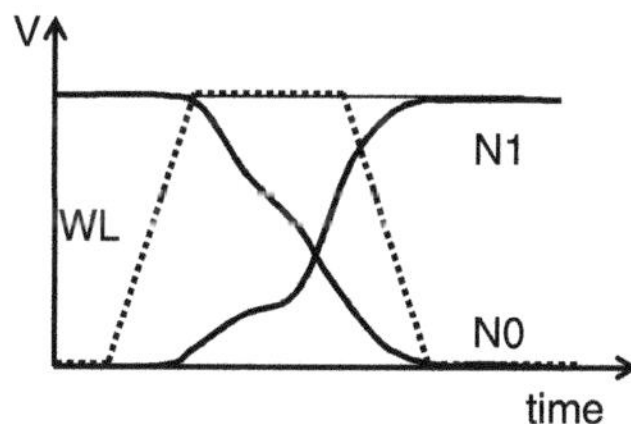

Fig. 3.8 Voltage waveforms of SRAM cell nodes during normal write operation

In the SRAM cell, stabilizing read operation and that of write operation have conflicting relationship. For example, the methods to stabilize the read and write operation are shown in Table 3.1. Therefore, it is difficult to improve stability of both read and write operations simultaneously. In all of the conventional manufacturing process before 130 nm process, both operations could be stabilized without problem as there were large operating margins in the SRAM cell. However, in recent

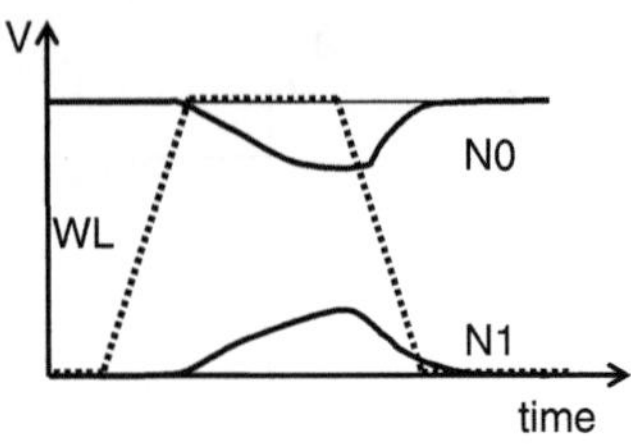

Fig. 3.9 Voltage waveforms of SRAM cell nodes during failure write operation

Table 3.1 Drivability change to enhance stability of each operation

	Access Tr	Driver Tr	Load Tr
Read	Small	–	Large
Write	Large	–	Small

manufacturing process, with increasing V_{th} variation of MOSFETs, the operating margin has been decreased as MOSFET current in SRAM memory cell is fluctuated by the V_{th} variation. Besides, the V_{th} variation destroys MOSFET current balance between read and write operations. Therefore, in SRAM cell design phase, to maximize SRAM operating margins, both read and write margins, is indispensable. The V_{th} is one of the good performance indicators of MOSFET, because the V_{th} can totally indicate the transistor performance. To realize this operating-margins maximization, precise analysis of SRAM operating margins is required. In this chapter, SRAM operating margin analyses that are influenced by MOSFET V_{th} variation are proposed.

3.2 V_{th} Window Curve

In SRAM design, V_{th} variation of transistors has critical influence on SRAM operation. V_{th} variations can be divided into local and global components. The local component is called as intra-die V_{th} variation, and is mainly due to random dopant fluctuation (RDF) in the channel and of line edge roughness (LER). The dopant distribution is increasing along the process scaling to control the short channel effect. The extent of variation is usually indicated as standard deviation. The standard deviation of local V_{th} variation is to inverse proportion to the square root of its gate length and its gate width [3.1]. On the other hand, the global variation is called die-to-die V_{th} variation. The global variation is due to fluctuation of MOSFET's gate length (L_g), gate width (W), gate oxide thickness (T_{ox}), and channel dopant concentration (N_a). The total variation is given as below. The process corners, for example FF or SS, are caused by these global components. V_{th} window analysis that analyzes SRAM operating margins is proposed. The V_{th} window analysis can be used when the SRAM cell design phase. As mentioned in the previous section, SRAM cell characteristic such as the SNM is quite sensitive to the V_{th} variability. In particular, a local V_{th} variation enhances the asymmetry of the

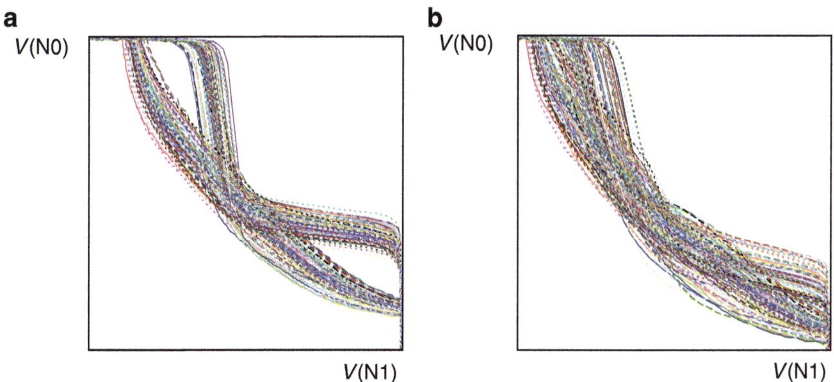

Fig. 3.10 Measured butterfly curves of prototype SRAM cells. (**a**) Butterfly curves without failure cells. (**b**) Butterfly curves with failure cells

pair MOSFET characteristic in a unit cell, leading to an enormous degradation of the functionality of the large-scale SRAM macro. The effect of local V_{th} variation is shown in Fig. 3.10. In this figure, measured butterfly curves of 45-nm prototype SRAM array are plotted. Some cells have sufficient SNM; on the other hand, some cells have insufficient SNM. The design of all cells is common, and therefore the SNM difference is caused by the local V_{th} variation. To analyze the SRAM cell stability, read and write limit lines are used.

Figure 3.11 shows the read and write limit lines. The horizontal axis is the central value of nMOS transistors in SRAM cell (DR0, DR1, AC0, and AC1 in Fig. 3.1), and the vertical axis is the central value of pMOS transistors in SRAM cell (LD0 and LD1 in Fig. 3.1). The solid line indicates operational limit line imposed by the static-noise margin: read margin (SNM). Generally, SNM is deteriorated by lower V_{th} of nMOS devices and higher V_{th} of pMOS devices. Therefore, in the left side and upper side of this SNM-limit line, retained data are destroyed by read operation, and consequently its read operation is to fail. If the SNM is larger, the line moves upper and left-hand direction. The dashed line indicates operational limits for write operation. Contrary to read operation, the writing margin is deteriorated by higher V_{th} of nMOS devices and lower V_{th} of pMOS devices. Therefore, in right-side and lower-side region of this write limit line, new data are not successfully written to the memory cell, which leads to write operation failure. If the write stability is larger, the line moves lower and right-hand direction. These operational limit lines are drawn by worst-case SNM and write operation analysis, which are simulated considering local V_{th} variations. The local V_{th} variation is added to the six transistors in SRAM cell individually. The worst pattern of V_{th} variation distribution to six transistors is precisely calculated by conventional method [3.2, 3.3]. These operational limit lines are called as V_{th} curves.

Figure 3.12 shows a V_{th} operating window, which is used in V_{th} window analysis. The V_{th} operating window is composed of the read and write limit lines. The

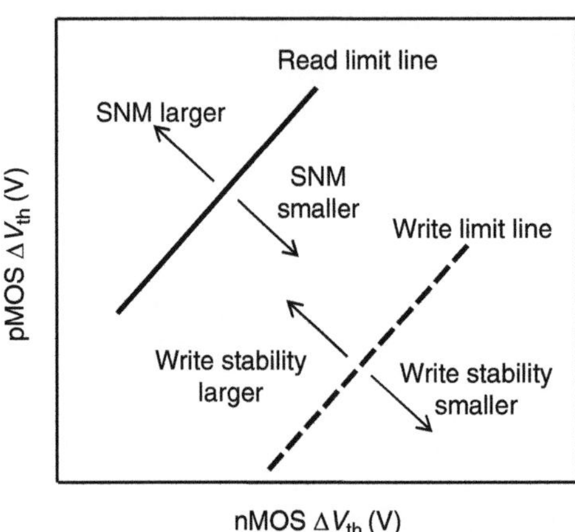

Fig. 3.11 Read and write limit lines (V_{th} curves)

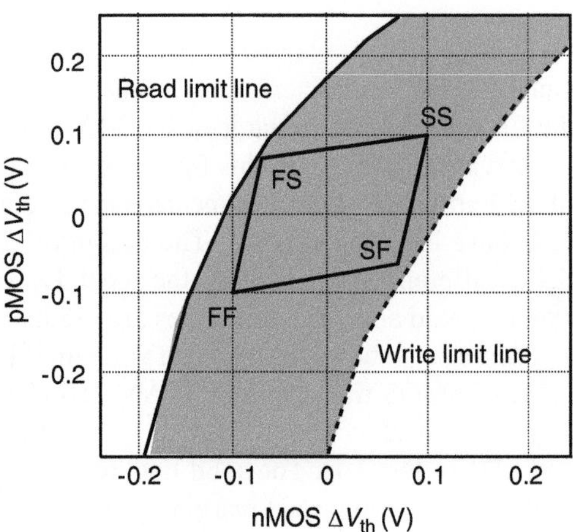

Fig. 3.12 V_{th} operating window

horizontal axis is the central value of nMOS transistors in SRAM cell, and the vertical axis is the central value of pMOS transistors in SRAM cell. These V_{th} values contain the global V_{th} variation. The diamond shape in Fig. 3.12 indicates the process corners. The FS point, for example, means nMOS fast and pMOS slow device.

When the V_{th} of manufactured SRAM transistors is in the window between each V_{th} curves (hatched region in Fig. 3.12), SRAM memory cell can operate correctly. Therefore, the diamond shape, which indicates the V_{th} distribution of manufactured devices, has to be within the area to guarantee SRAM correct operation in all process corners. This V_{th} window analysis is used in memory cell design, and the central V_{th} value (TT device) is adjusted so that V_{th} corners are within the operating window. An example of this V_{th} window analysis is described as below. The description supposes that an SoC is manufactured on the point of a read-limit line, and average value of nMOS V_{th} is -0.06 V, and that of pMOS V_{th} is 0.1 V. These values are

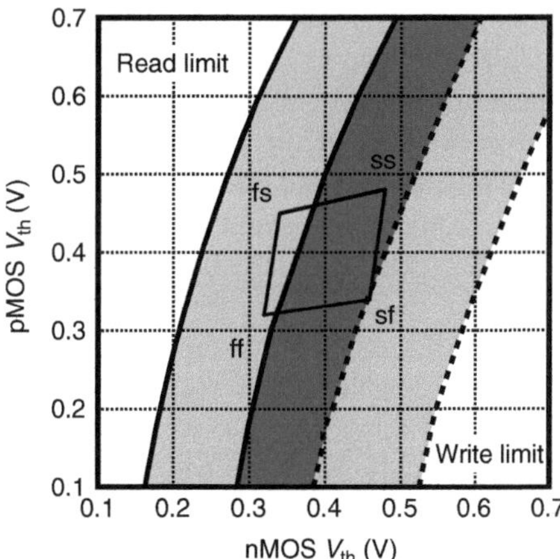

Fig. 3.13 V_{th} window when considering with 20 mV-s and 40 mV-s V_{th} variation

decided by the global V_{th} variation. In the SoC, there is local V_{th} variation, and when local V_{th} variation brings a lower V_{th} for access MOS in SRAM cell and a higher V_{th} for load MOS, read margin is deteriorated. Moreover, when the local V_{th} variation is maximized, the read margin becomes zero, and the operation is just correctly completed. If the global V_{th} variation brings slightly lower nMOS V_{th} or slightly higher pMOS V_{th}, read margin becomes minus, and the read operation is to fail. In short, on the read-limit line, the read operation is just correctly completed, and if there are more local V_{th} variation or much harder operating conditions, the operation is to fail. Figure 3.13 shows the V_{th} window, when considering a 20 mV-standard deviation and 40 mV-standard deviation V_{th} variation [3.2]. The V_{th} variation becomes larger according to the advance of process scaling, and the V_{th} window also becomes narrower according to the advance of process scaling. In this V_{th} window analysis, the local V_{th} variation is considered in the simulation to plot the operational limit lines, and the global V_{th} variation is represented as the process corners indicated as a diamond shape in the V_{th} window analysis. Although V_{th} variation of local component is influenced by the global component, the relationship between these two components is not considered in this V_{th} window analysis. The more accurate V_{th} window analysis has to consider the influence of the global V_{th} variation to the local V_{th} variation.

3.3 Sensitivity Analysis

Now that we found the importance and the usefulness of the V_{th} window analysis in the advanced SRAM cell design, we then explain how the corresponding boundaries, V_{th} curves, are obtained. This methodology investigates the worst-case model that minimizes read or write operation. The merit to use the worst-case model is that we can obtain simulation results much faster than the well-known Monte Carlo simulation (MC). In the MC simulation, all the possible V_{th} combinations within a unit SRAM cell are generated, and the corresponding margins are all calculated. In particular for the V_{th} window analysis, we need to take the large number of SRAM cell array (e.g., 1 Mbits or larger) into consideration depending on the capacity of the actual SRAM products. This means that more than 10^6 simulation times would be needed to identify single point of V_{th} curves. In contrast, our method requires about 50-time simulations for each point because this method is based on the sensitivity analysis and searches for the worst V_{th} combination that minimizes read or write margin: excessive simulations that do not contribute to the estimate of the boundary are not calculated. Furthermore, a slight modification of the simulation flow allows us to estimate the minimum operating voltage of the SRAM cell array (V_{ddmin}), which is also important in characterizing macroscopic feature of a large-scale SRAM array. The validity of using the worst-case model is verified by comparing with the measurement result.

We start our discussion by expanding SRAM DC margin. Hereafter, we only focus on the read margin or the SNM, which can be also extended to the write margin. Although various factors such as supply voltage and body bias effects affect the SNM drastically, we simply assume that the SNM is a function of 6 transistor's V_{th}s' in a unit memory cell [3.4]. Furthermore, the sensitivity analysis shown in Fig. 3.14 indicates that 2 transistors in a memory cell have no contribution to the SNM, which allows us to denote the SNM function as:

$$\text{SNM}_L(Vt_{(AC1)}, Vt_{(DR0)}, Vt_{(DR1)}, Vt_{(LD0)}). \tag{3.1}$$

Since our task is to derive the situation of $\text{SNM} = 0$ due to the V_{th} variation, it is instructive to divide each $Vt_{(X)}$ into several components.

$$Vt_{(X)} = Vt_{(X\_typ)} + \Delta Vt_{(X)} + x\sigma Vt_{(X)}, \tag{3.2}$$

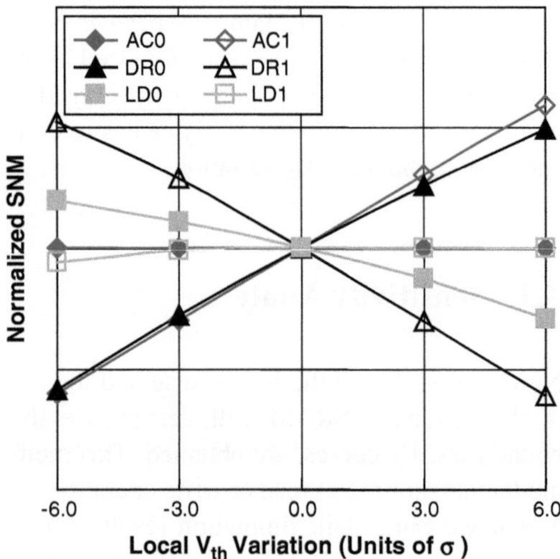

Fig. 3.14 Sensitivity analysis

where $Vt_{(X\_typ)}$ is the typical V_{th} value of transistor X, $\Delta Vt_{(X)}$ the deviation from the typical V_{th} which is equivalent to the global variation, and $\sigma Vt_{(X)}$ the standard deviation of local V_{th} variation. As mentioned in the previous subsection, $\sigma Vt_{(X)}$ can be calculated if we define the transistor dimension L/W as well as a proportionality coefficient, Avt [3.1, 3.5]. In addition, the parameter x indicates the probability of appearance, which plays an important role in our later discussion in determining the worst-case model for SNM.

Based on the Taylor expansion, we expand (3.1) around the point of $Vt_{(X\_typ)} + \Delta Vt_{(X)}$ for each transistor,

$$SNM_L(Vt_{(AC1)}, Vt_{(DR0)}, Vt_{(DR1)}, Vt_{(LD0)})$$

$$= SNM_C + \frac{\partial SNM}{\partial(x \times \sigma Vt_{(AC1)})} x \times \sigma Vt_{(AC1)} + \frac{\partial SNM}{\partial(y \times \sigma Vt_{(DR0)})} y \times \sigma Vt_{(DR0)}$$

$$+ \frac{\partial SNM}{\partial(z \times \sigma Vt_{(DR1)})} z \times \sigma Vt_{(DR1)} + \frac{\partial SNM}{\partial(t \times \sigma Vt_{(LD0)})} t \times \sigma Vt_{(LD0)}, \qquad (3.3)$$

where the SNM_C indicates the value at $Vt_{(X\_typ)} + \Delta Vt_{(X)}$, meaning the SNM without local variation. The factor $\partial SNM/\partial(\sigma Vt)$ corresponds to the gradient appeared in the sensitivity analysis (see Fig. 3.14). Furthermore, coefficients (x, y, z, t) express the combination of probability for each local variation to appear, which correlates with a failure appearance probability out of the total memory capacitance, r:

$$x^2 + y^2 + z^2 + t^2 = r^2. \qquad (3.4)$$

For instance, we should take $r = 6$ restriction into consideration in designing 505 Mbit SRAM array in one chip. This equation implies that each component is assumed to be independent (or orthogonal), and that (x, y, z, t) corresponds to a vector on the sphere with radius r. Note that our purpose is to find out the point where SNM becomes 0 by changing $\Delta Vt_{(X)}$ (global deviation from typical V_{th} condition) within the restriction of equation (3.4). For the sake of a latter mathematical treatment, we introduce the following four-dimensional spherical expressions in advance.

$$x = r \sin \xi \sin \phi \cos \theta$$

$$y = r \sin \xi \sin \phi \sin \theta$$

$$z = r \sin \xi \cos \phi$$

$$t = r \cos \xi. \qquad (3.5)$$

We then move on to the next step of solving (3.3) by setting $SNM_L = 0$. Mathematically, it should be required that the infinitesimal deviation of SNM against the angle parameter (ξ, φ, θ) be 0 where the SNM undergoes the smallest value on the sphere of radius r, leading to:

$$\frac{\partial \text{SNM}}{\partial \xi} = \frac{\partial \text{SNM}}{\partial \phi} = \frac{\partial \text{SNM}}{\partial \theta} = 0. \tag{3.6}$$

By solving this equation, we can straightforwardly obtain the following solutions:

$$x^{(1)} = -\frac{r}{R^{(0)}} \frac{\partial \text{SNM}}{\partial (x^{(0)} \sigma Vt)} \times \sigma Vt_{(AC1)}$$

$$y^{(1)} = -\frac{r}{R^{(0)}} \frac{\partial \text{SNM}}{\partial (y^{(0)} \sigma Vt)} \times \sigma Vt_{(DR0)}$$

$$z^{(1)} = -\frac{r}{R^{(0)}} \frac{\partial \text{SNM}}{\partial (z^{(0)} \sigma Vt)} \times \sigma Vt_{(DR1)}$$

$$t^{(1)} = -\frac{r}{R^{(0)}} \frac{\partial \text{SNM}}{\partial (t^{(0)} \sigma Vt)} \times \sigma Vt_{(LD0)}, \tag{3.7}$$

where,

$$R^{(0)} = \sqrt{\left(\frac{\partial \text{SNM}}{\partial (x^{(0)} \sigma Vt)} \sigma Vt_{(AC1)}\right)^2 + \left(\frac{\partial \text{SNM}}{\partial (y^{(0)} \sigma Vt)} \sigma Vt_{(DR0)}\right)^2 + \left(\frac{\partial \text{SNM}}{\partial (z^{(0)} \sigma Vt)} \sigma Vt_{(DR1)}\right)^2 + \left(\frac{\partial \text{SNM}}{\partial (t^{(0)} \sigma Vt)} \sigma Vt_{(LD0)}\right)^2}.$$

Since the worst vector that minimizes the SNM is derived iteratively, the suffix (0) in (3.7) indicates the initial worst vector chosen arbitrarily. In this way, one can obtain analytically the combination of local variation that minimizes SNM. Subsequently, to search for the point where the SNM becomes 0, we need further examination by substituting the worst vector obtained above and calculating corresponding SNM. Our aim is to know the $\Delta Vt_{(X)}$ condition for both NMOS and PMOS, which enables us to draw a curve on the V_{th} coordinate. According to the flow chart shown in Fig. 3.15, one can reach the V_{th} curve for SNM.

Further understanding of this methodology allows us to estimate the experimental data using the actual SRAM test chip. In deriving V_{th} curves, the failure rate r was recognized as a given parameter determined by the design of the SRAM capacitance. As mentioned above, we can visually estimate the extent of the global variation that ensures a stable DC function for both read and write operation. On the contrary, logically speaking, if we have a certain fixed value of V_{th} (or the amount of deviation from the typical V_{th} condition denoted by a SPICE parameter), the r value can be derived. This indicates that if we have an averaged V_{th} value of each SRAM transistor in an actual test chip, we can estimate the fail bit count (FBC) rate dependence on the supply voltage in the SRAM probe test. In other words, this methodology enables us to evaluate the minimum supply voltage (V_{ddmin}) of the SRAM Macro, which is one of the most important characteristics of SRAM cell array from the macroscopic point of view.

In Fig. 3.16, we show the experimental FBC dependence on the supply voltage V_{DD} as well as the simulated results. Two dotted lines indicate the experimental

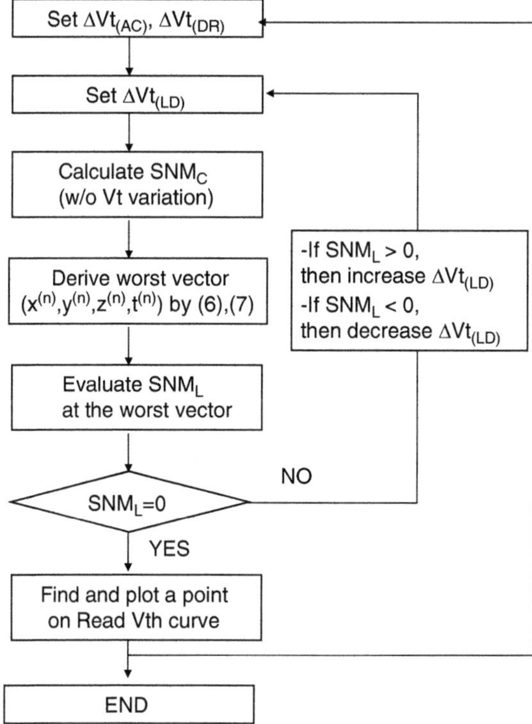

Fig. 3.15 V_{th} analysis flow chart

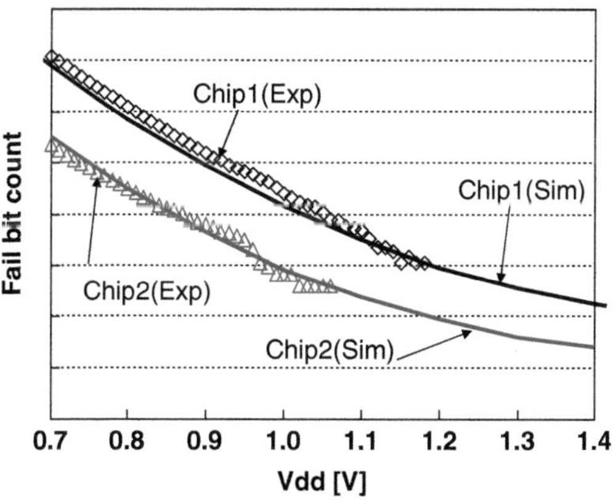

Fig. 3.16 Experimental FBC dependence on supply voltage

results evaluated from the different chips on different wafers. The SRAM cell ($0.499\,\mu m^2$ [3.6]) was fabricated on our original process of 65 nm bulk CMOS technology. For these two chips we examined, the PMOS V_{th}s were realized much higher than our target, so that the read (SNM) FBC has a relatively lower dependence on the V_{DD} change. To compare the experiments with the simulation results based on our methodology, we evaluated V_{th} and the corresponding local variation of each SRAM transistor using the other test circuitry. It is obvious that the simulated results are in good accordance with the experiment, exemplifying the validity of our methodology.

References

3.1. M.J. Pelgrom, A.C.J. Duinmaijar, Matching properties of MOS transistors. IEEE. J. Solid-State Circuits **24**(5), 1433–1440 (1989)
3.2. M. Yamaoka et. al., Low power SRAM menu for SOC application using Yin-Yang-feedback memory cell, in *Proceedings of 2004 Symposium on VLSI Circuits*, pp. 288–291
3.3. Y. Tsukamoto, K. Nii, S. Imaoka, Y. Oda, S. Ohbayashi, T. Yoshizawa, H. Makino, K. Ishibashi, H. Shinohara, Worst-case analysis to obtain stable read/write DC margin of high density 6T-SRAM-array with local Vth variability, in *Proceedings of ICCAD*, Nov. 2005, pp. 394–405
3.4. F. Tachibana, T. Hiramoto, Re-examination of impact of intrinsic dopant fluctuations for ultra-small bulk and SOI CMOS, IEEE Trans. Electron Devices **48**(9), 1995–2001 (2001)
3.5. P.A. Stolk, F.P. Widdershoven, D.B.M. Klaassen, Modeling statistical dopant fluctuations in mos transistors IEEE Trans. Electron Devices **45**(9), 960–1971 (1998)
3.6. S. Ohbayashi et al., A 65 nm SoC embedded 6T-SRAM design for manufacturing with read and write cell stabilizing circuits, in *Proceedings of 2006 Symposium on VLSI Circuits*, 2006, pp. 17-18

Chapter 4
Low Power Memory Cell Design Technique

Kenichi Osada and Masanao Yamaoka

Abstract This chapter describes the low power memory cell design technique. Section 4.1 introduces fundamentals of leakage of SRAM array. In Sect. 4.2, source line voltage control techniques are explained as new designs to reduce standby power dissipation. Using the techniques for a 16-Mbit SRAM chip fabricated in 0.13-μm CMOS technology, the cell-standby current is 16.7 fA at 25°C and 101.7 fA at 90°C. By applying the techniques to 1-Mbit 130-nm embedded SRAM, the leakage current is reduced by about 90% from 230 to 25 μA. In Sect. 4.3, a new SRAM cell layout design developed for low-voltage operation is described. A lithographically symmetrical cell for lower-voltage operation was developed. The measured butterfly curves indicate that the memory cell has a large enough noise margin even at 0.3 V.

4.1 Fundamentals of Leakage of SRAM Array

DRAM cell has low standby power dissipation [4.1]. But it needs refresh treatment. On the other hand, SRAM cell does not need special treatment like refresh because the full CMOS 6-T memory cell statically retains data. Hence, it is easy to control SRAM chip, which is used in mobile phones. Moreover, SRAM is also used as the embedded memory in the application processor in mobile cellular phones since it is fabricated by the same process as CMOS logic uses. Advances in MOS technology are accompanied by increases in the gate-oxide tunnel leakage and

K. Osada (✉)
Measurement Systems Research Department, Central Research Laboratory, Hitachi Ltd.,
1-280, Higashi-koigakubo, Kokubunji-shi, Tokyo 180-8601, Japan
e-mail: kenichi.osada.aj@hitachi.com

M. Yamaoka
1101 Kitchawan Road, Yorktown Heights, NY 10598, USA
e-mail: masanao.yamaoka@hal.hitachi.com

K. Ishibashi and K. Osada (eds.), *Low Power and Reliable SRAM Memory Cell and Array Design*, Springer Series in Advanced Microelectronics 31, DOI 10.1007/978-3-642-19568-6_4, © Springer-Verlag Berlin Heidelberg 2011

gate-induced drain leakage (GIDL) currents, which are added to the subthreshold leakage current as the main components of leakage [4.2–4.4]. The increased leakage currents increases leakage current of SRAM cell. In this section, the fundamentals of leakage power of SRAM array will be mentioned.

4.1.1 *Leakage Currents in an SRAM of Conventional Design*

The problem of leakage current increases as technology advances. The four significant leakage currents in a technologically advanced nMOS transistor are shown in Fig. 4.1. These are the gate-oxide tunnel leakage, the GIDL, the junction tunnel leakage, and the subthreshold leakage currents. Subthreshold leakage is a channel leakage current that flows from drain to source while the gate is off, since this state sets up a weak inversion layer in the channel region.

The significant leakage currents in an advanced technology SRAM cell of conventional design are shown in Fig. 4.2. In the conventional SRAM, 1.5 V is applied to the bit lines (BT and BB) and power line (V_{DDI}), and 0 V is applied to the word line and the V_{SS} line (V_{SSM}). In the cell as a whole, two gate-tunnel leakage currents, five GIDL currents, and three subthreshold leakage currents constitute the main leakage currents. The use of high threshold voltages provides a straightforward way to reduce the subthreshold leakage currents although the cell current for read operation decreases. However, it is not easy to reduce the tunnel leakage and GIDL currents by simply improving the device structure.

4.1.2 *Gate-Tunnel Leakage and GIDL Currents*

In this subsection, we describe the gate-tunnel leakage and GIDL currents. Measured plots of gate-tunnel leakage current as a function of gate voltage with the thickness of the gate oxide (T_{OX}) varied as a parameter are given in Fig. 4.3. With a 2-Å decrease in T_{OX}, there is a tenfold increase in gate-tunnel leakage current. As

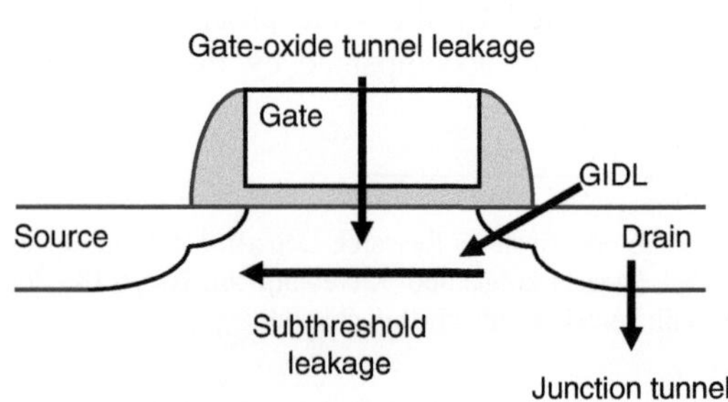

Fig. 4.1 Leakage currents in a technologically advanced nMOS device

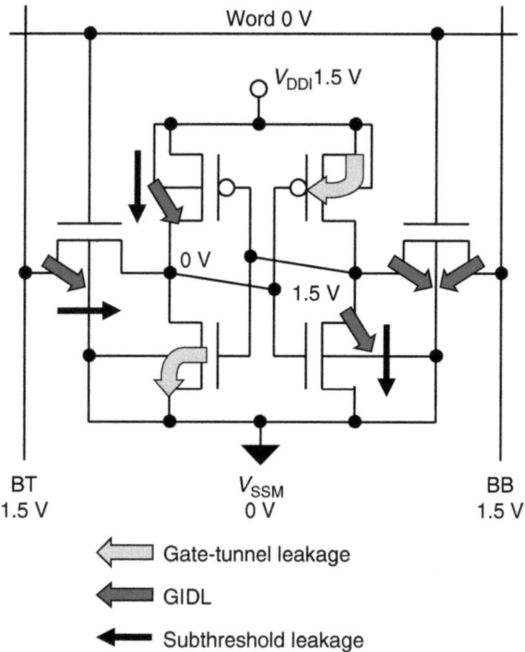

Fig. 4.2 Standby leakage currents in an advanced technology SRAM cell of conventional design

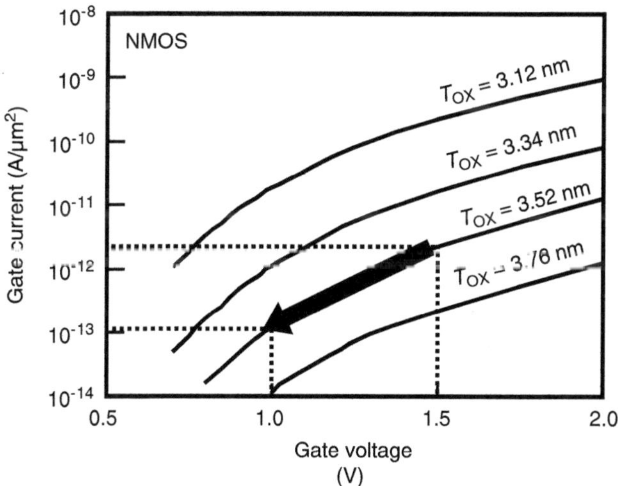

Fig. 4.3 Gate-tunnel leakage current as a function of gate voltage with the thickness of the gate oxide (T_{ox}) varied as a parameter

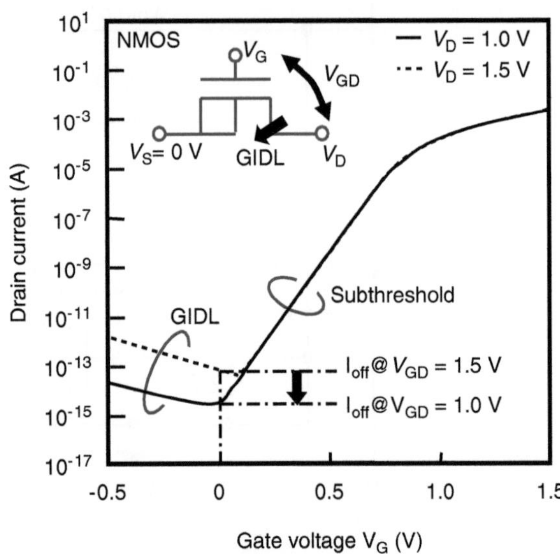

Fig. 4.4 GIDL and subthreshold leakage currents

technology advances, this current becomes the main form of leakage. On the other hand, the gate current decreases with the gate voltage. Reducing the voltage by 0.5 V (from 1.5 to 1.0 V) reduces this leakage component by 95%. This is because the tunnel leakage current simply decreases with the electric field strength in the gate oxide [4.2–4.6].

Figure 4.4 shows measured curves of drain current as a function of the gate voltage (V_G). The solid line is the drain current when the drain voltage (V_D) is 1.0 V, and the dotted line is the drain current when V_D is 1.5 V. The subthreshold leakage currents are dominant in the region where V_G is greater than 0 V, while the GIDL currents are dominant in the region where V_G is less than 0 V. Since a raised threshold voltage is in use, the subthreshold leakage currents at $V_G = 0$ V are negligibly weak in comparison with the GIDL currents. The standby current is thus almost equivalent to the GIDL current. The GIDL current (I_{GIDL}) is very sensitive to the electric field F and is given by [4.4, 4.7]:

$$I_{GIDL} = AF5/2 \exp(-B/F), \tag{4.1}$$

where A and B depend on the bandgap EG and constants. This expression indicates that the logarithm of the GIDL current is in inverse linear proportion to the gate voltage, which is mainly responsible for the electric field directly beneath the gate oxide. That is, the amount of GIDL current is determined by the difference between the voltages at the gate and drain (V_{GD}). If V_{GD} is reduced from 1.5 to 1.0 V, the electric field strength is relaxed and the GIDL current in our device is reduced by about 90%. If the threshold voltage is low, the main part of the leakage currents is the subthreshold leakage current. The subthreshold leakage current is reduced by the applications of a negative V_{GS} or a negative V_{BB} back-body bias voltage.

4.2   Source Line Voltage Control Technique

In this section, we describe source line voltage control techniques as new designs to achieve the lowest-ever [4.8–4.10] levels of standby power dissipation. Conventionally, the power line voltage is constant to guarantee stable operations of SRAM circuit. Here, the power line voltage is controlled to improve its leakage performance. At first, electric field relaxation (EFR) scheme is described as the technique for low power SRAM chip used in mobile phones. Higher threshold voltage is used in SRAM cell for low standby power. Second, we describe the leakage reduction techniques which are used in embedded SRAM modules in the application processor in mobile cellular phones. Lower threshold voltage is used in SRAM cell for low access time.

4.2.1   EFR Scheme for Low Power SRAM

The EFR scheme has thus been developed as a way of attacking these leakage currents at the circuit level. The EFR scheme is depicted in Fig. 4.5. At retention mode, a 1.0 V supply replaces the 1.5-V supply to the bit lines (BT and BB). V_{SSM} is raised from 0 to 0.5 V. The black solid arrows indicate differences in potential that are relaxed from 1.5 to 1.0 V. This relaxation reduces the gate leakage and GIDL currents by about 90%. The solid gray arrow indicates the application of a negative 0.5-V V_{GS} to the transfer nMOS transistor. This negative bias brings the subthreshold leakage current close to zero. The dotted arrow indicates a V_{BB} back-body bias voltage of negative 0.5 V that is applied to the driver nMOS transistor. This back bias reduces the subthreshold leakage current by about 90%. The difference between the voltages of the source and the substrate increases from 0 to 0.5 V. However, the band-to-band tunneling leakage is still negligible.

Figure 4.6 gives measured standby leakage currents at 25°C (Fig. 4.6a) and 90°C (Fig. 4.6b) for SRAM cells of the conventional design and with EFR, both of which were produced under worst-case process conditions. The subthreshold leakage and GIDL components have been summed for each device type. In both cases, the pMOS threshold voltage is 1.0 V and there is much less subthreshold leakage than GIDL current at both 25°C and 90°C. The nMOS threshold voltage is 0.7 V and the subthreshold leakage current is smaller at 25°C but larger at 90°C than the GIDL current. Values for cell-standby leakage current in the cell with the EFR scheme are 16.6 fA at 25°C and 101.7 fA at 90°C, 82.5% and 91.8% smaller, respectively, than the values for the conventional cell.

4.2.2   Chip Architecture

Figure 4.7 shows a circuit diagram of one mat of the fabricated SRAM chip, which consists of 32 mats. Each mat is composed of 2,048 word lines by 256-data-bit and 20-parity-bit columns. The mat is divided into four banks. The timing diagram for a

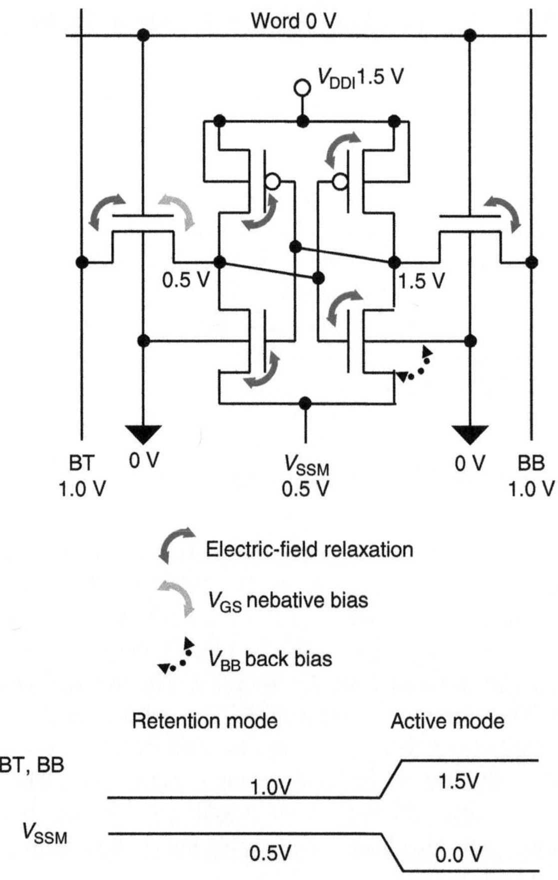

Fig. 4.5 Electric field relaxation scheme

simulated read operation is shown in Fig. 4.8. This SRAM is asynchronous. SRAM is in retention mode unless an address signal changes. If an address signal changes, SRAM is shifted from retention mode to active mode. An ATD pulse is generated and actives the reset operation. During the 4.3-ns reset period, the memory cell V_{SS} line (V_{SSM}) of the selected bank is pulled from 0.5 to 0 V, while the bit-line pairs (BT, and BB) of the mat are precharged from 1.0 to 1.5 V. The conventional asynchronous SRAM always has the reset time for the bit-line pair precharge from 0 to 1.5 V, and we simply used the reset time for bit-line pairs precharge and V_{SSM} discharge in our EFR scheme. Therefore, the access time does not increase because of the EFR scheme as the asynchronous SRAM. A word line (wl) is then activated. With activation of the sense amplifiers, the selected 128 bits of data and 10 parity bits appear on the local bus (LBUS). To reduce the introduction of noise to the power supplies, V_{SSM} is controlled at the bank level instead of at the mat level. This form of control keeps the V_{SS} supply's noise level at a low 32 mV. This result was obtained in simulation through the use of power supply nets. The V_{SSM} line is reinforced with the third metal to accommodate the V_{SSM} control from 0.5 to 0 V. The V_{SSM} line

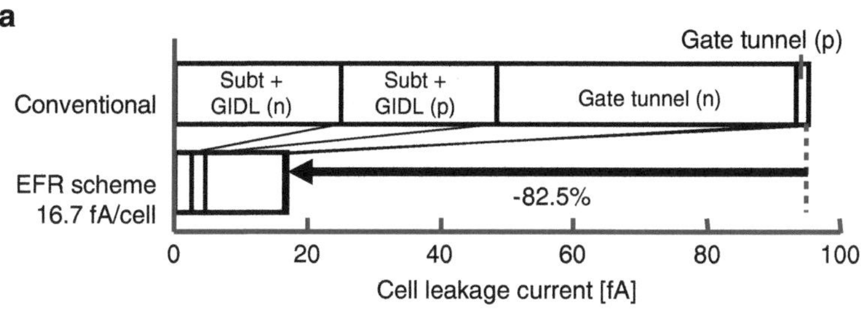

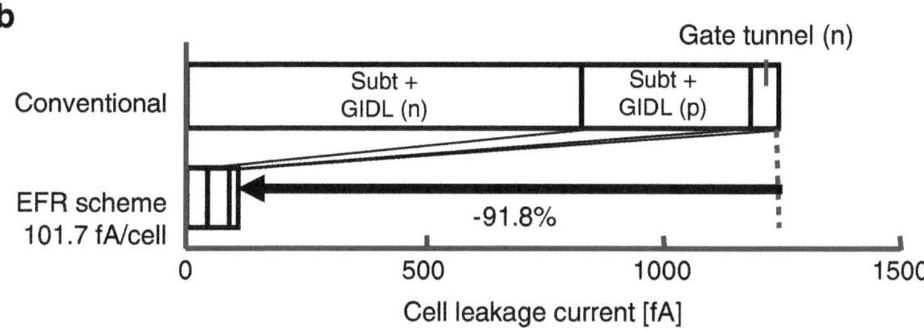

Fig. 4.6 Measured SRAM cell leakage current; (**a**) 25°C/worst case (**b**) 90°C/worst case

is very slowly recovered back to the 0.5 V level after the completion of the read operation in 500 μs. The bit-line pairs are simply recovered back to the 1.0 V level after the completion of the read operation by the leakage in 100 ms. Therefore, the reset operation hardly affects the dynamic power consumption.

4.2.3 *Results*

This SRAM was fabricated using 3-metal 0.13-μm CMOS technology. Figure 4.9 shows a photograph of the fabricated test chip [4.11]. The process and device parameters and the chip's features are listed in Fig. 4.9. The gate length of the NMOS and PMOS devices is 0.14 μm. The threshold voltages of the memory cells are 0.7 V for NMOS and 1 V for PMOS. On the other hand, the threshold voltage of the peripheral circuits is 0.3 V. The gate-oxide layers of the internal circuits are 3.7-nm thick and those of the external circuits are 8.4-nm thick. The cell size is 0.92 by 2.24 μm. The gate width of the transfer NMOS and the load PMOS devices is 0.18 μm. The gate width of the driver NMOS device is 0.24 μm. We used the same sizes as those of the conventional SRAM cell. The test-chip access time at 3.3 V is 27 ns. At 25°C, the cell leakage current is 16.7 fA, the chip leakage is 0.5 μA, and the voltage-converter leakage is 0.13 μA. The access time is 27 ns. Power dissipation is 19 mW with a 70-ns cycle.

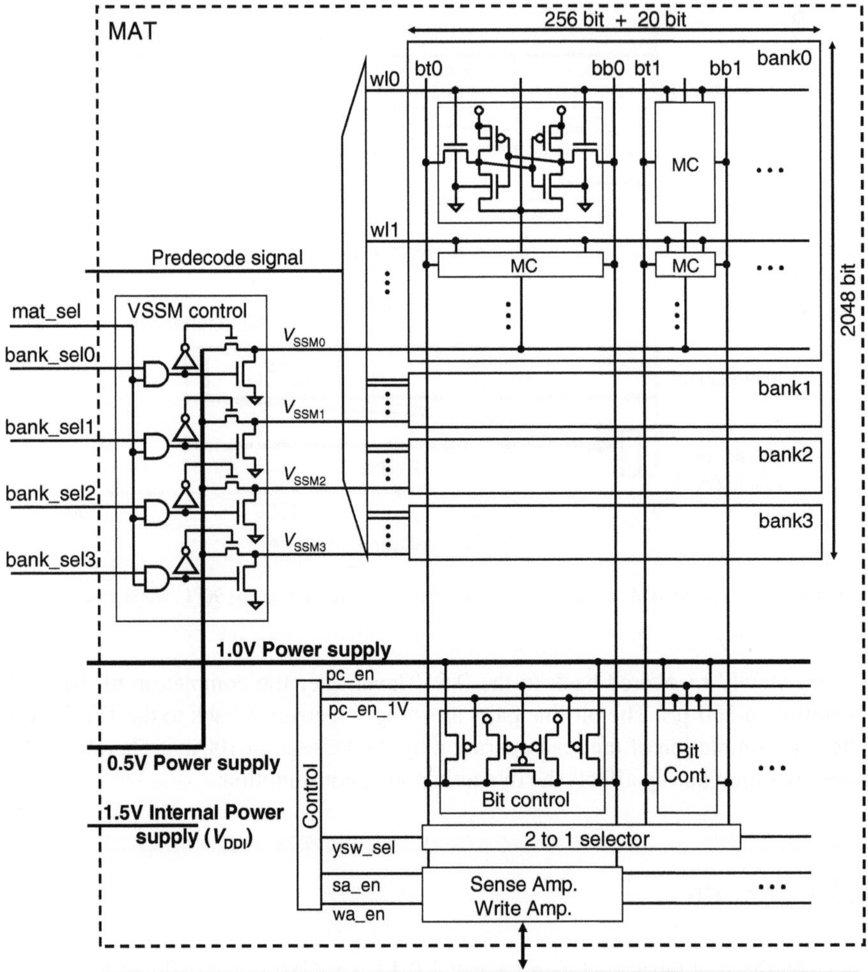

Fig. 4.7 Circuit diagram of one mat

4.2.4 Source Line Voltage Control Technique for SRAM Embedded in the Application Processor

The leakage reduction techniques described in this section focus on embedding in the application processor in mobile cellular phones. In general, for SRAM embedded in the application processor, the lower threshold voltage is used. Therefore, the main part of the leakage currents is the subthreshold leakage current. To reduce the leakage current of SRAM cells during non-accessed state, source line

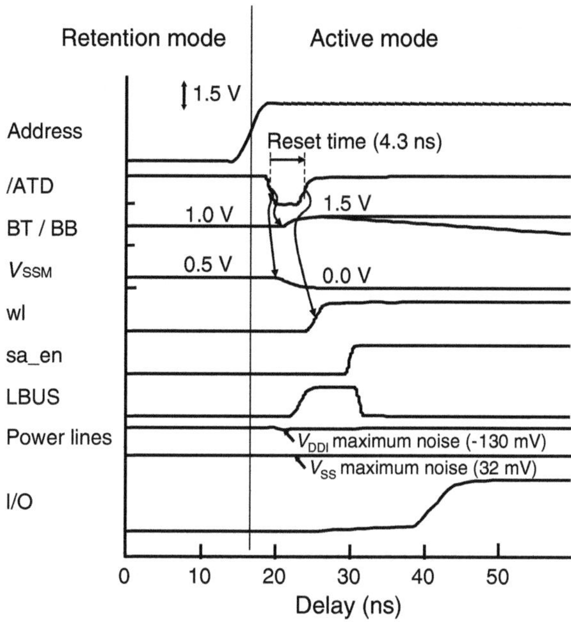

Fig. 4.8 Simulated waveforms for read operation

voltage control techniques are proposed. The proposed techniques control power line voltage with minimum sacrifice of its operation stability.

The power supply voltage of SRAM cells is proposed to be controlled to intermediate voltage between V_{DD} and V_{SS}, and the intermediate voltage enables both leakage reduction and data retention. The intermediate voltage has to be generated by simple circuit, because if a complex circuit is used, extra power consumption and an area penalty are necessary. This section proposes a simple voltage control circuit, which is composed of only power switch, resistance, and diode. The circuit can eliminate extra power consumption and an area penalty.

The SRAM module must satisfy its performance target, and it is designed to be stable enough to retain its data. When controlling V_{SSM} voltage to reduce memory cell leakage, the higher V_{SSM} reduces SRAM cell leakage current to less level. On the other hand, it deteriorates SRAM retention stability. This relationship is indicated in Fig. 4.10. To satisfy both low leakage and high stability, the V_{SSM} have to be controlled to the lowest voltage that satisfies the leakage target. Figure 4.11 shows a V_{SSM} controller, PLVC1, which consists of three NMOSFETs. One NMOSFET works as a power switch (MS1) between V_{SSM} and V_{SS}, one as a diode (MD1), and the other as a resistor (MR1). MR1 has a long gate and is normally on. The main leakage current of a memory cell produced by the 130-nm manufacturing process is subthreshold leakage, which is largely influenced by V_{th} fluctuation. For example, if V_{th} decreases by 100 mV, the subthreshold leakage increases by more than ten times; on the other hand, a drive current of a MOSFET increases by only

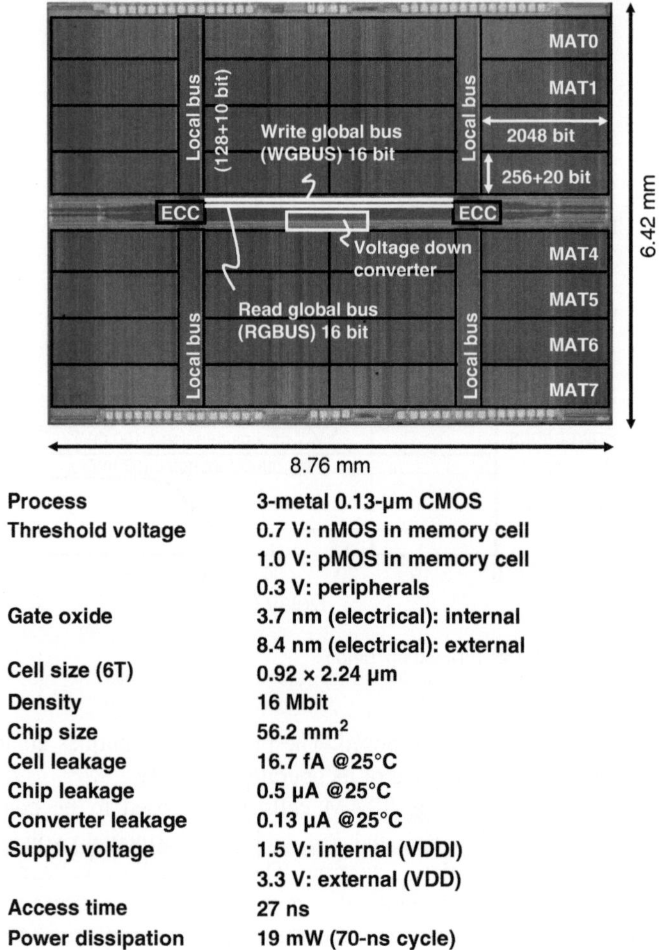

Process	3-metal 0.13-µm CMOS
Threshold voltage	0.7 V: nMOS in memory cell
	1.0 V: pMOS in memory cell
	0.3 V: peripherals
Gate oxide	3.7 nm (electrical): internal
	8.4 nm (electrical): external
Cell size (6T)	0.92 × 2.24 µm
Density	16 Mbit
Chip size	56.2 mm^2
Cell leakage	16.7 fA @25°C
Chip leakage	0.5 µA @25°C
Converter leakage	0.13 µA @25°C
Supply voltage	1.5 V: internal (VDDI)
	3.3 V: external (VDD)
Access time	27 ns
Power dissipation	19 mW (70-ns cycle)

Fig. 4.9 Photograph and features of the chip

about 10%. When the V_{th} of manufactured MOSFET is fluctuated as high value, the memory cell leakage largely decreases. On the other hand, the current through MR1 slightly decreases, because MR1 is in on state. If the MR1 current is greater than the memory cell leakage, the V_{SSM} voltage becomes low. On the other hand, when V_{th} is fluctuated to lower value, the V_{SSM} voltage is high, but MD1 restricts a rise in the V_{SSM} voltage. This keeps the voltage low enough to retain the stored data. Figure 4.12 shows V_{th} of manufactured MOSFET vs. V_{SSM} voltage. The horizontal axis shows the V_{th} difference between designed value and actual manufactured value, which varies by process variation. When V_{th} of manufactured transistor is low and the leakage current is high, the V_{SSM} voltage has to be high to significantly reduce the leakage current. When V_{th} of manufactured transistor is high and the

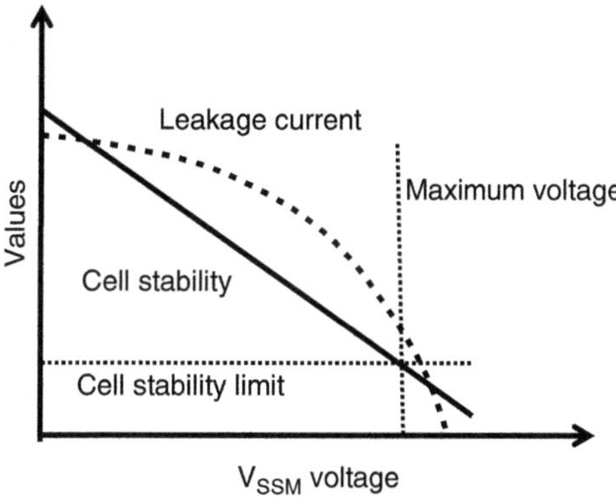

Fig. 4.10 Leakage and stability vs. local power line voltage

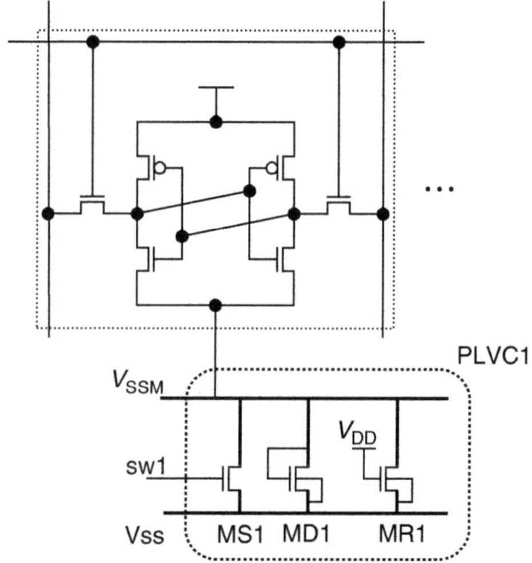

Fig. 4.11 V_{SSM} controller in embedded SRAM

leakage current is low, the V_{SSM} voltage has to be low to keep SRAM retention stability. The broken line indicates the ideal values that just satisfy the leakage target and maximize memory cell stability. If the MOSFETs are worst-leakage devices, their V_{th} of MOSFETs is 0.1 V lower than the designed value, and the V_{SSM} voltage has to be higher than 0.3 V to satisfy the leakage target. If the voltage controller is composed of only a power switch and a diode, V_{SSM} voltage is shown as line (a) in the graph. This line satisfies the leakage target, but the memory cell stability becomes low. If the voltage controller PLVC1 is composed of a constant-voltage source circuit, V_{SSM} voltage is shown as line (b) in the graph. This line satisfies the leakage target, and the stability is better than using line (a), but the memory cell stability is still low. Therefore, using three types of MOSFETs, MS1, MD1, and

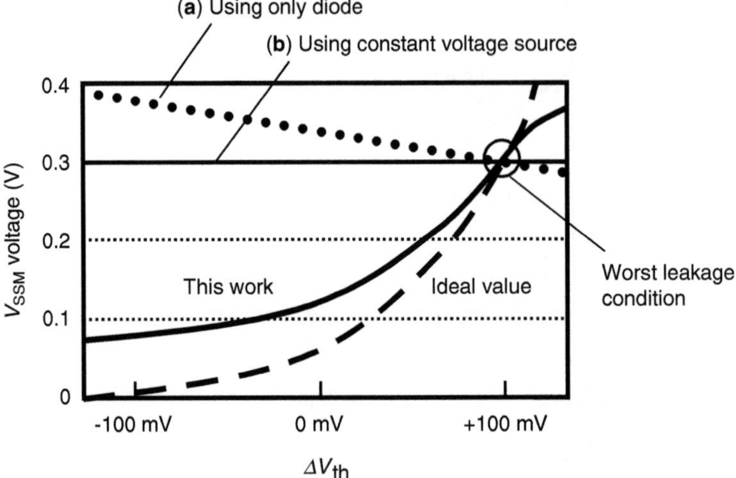

Fig. 4.12 V_{SSM} voltage when V_{th} is changed

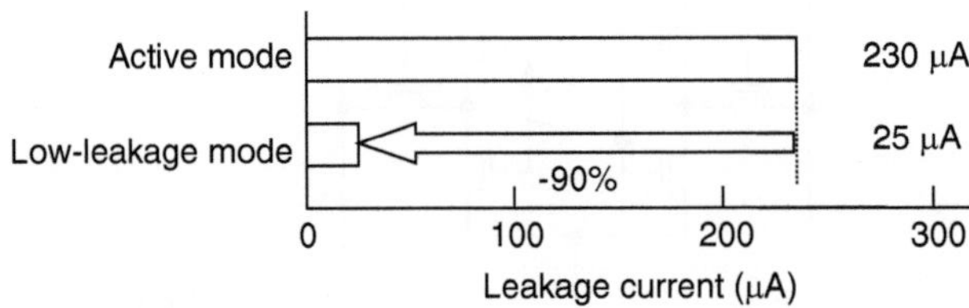

Fig. 4.13 Effect of leakage reduction in 1-Mbit leakage-worst embedded SRAM

MR1, for voltage controller, V_{SSM} voltage (bold line in Fig. 4.12) becomes close to the ideal value. These values satisfy the leakage target and keep the stability higher. The PLVC1 satisfies the leakage target and keeps the SRAM stability high, simultaneously.

In Fig. 4.13, leakage current of 1-Mbit prototype SRAM is indicated [4.12]. Using leakage reduction technique, the leakage current is reduced by about 90%. In Fig. 4.14, the chip photograph and features of the 130-nm prototype are indicated.

4.3 LS-Cell Design for Low-Voltage Operation

Low power dissipation microprocessors, which vary their operation frequency and power supply voltage according to system loads, have recently been developed for portable electronic devices [4.13]. Moreover, the next generation microprocessors need lower-voltage-range operation for lower power consumption. The on-chip cache of such microprocessors, however, acts as a bottleneck that blocks such performance improvements. A new SRAM cell layout design developed for

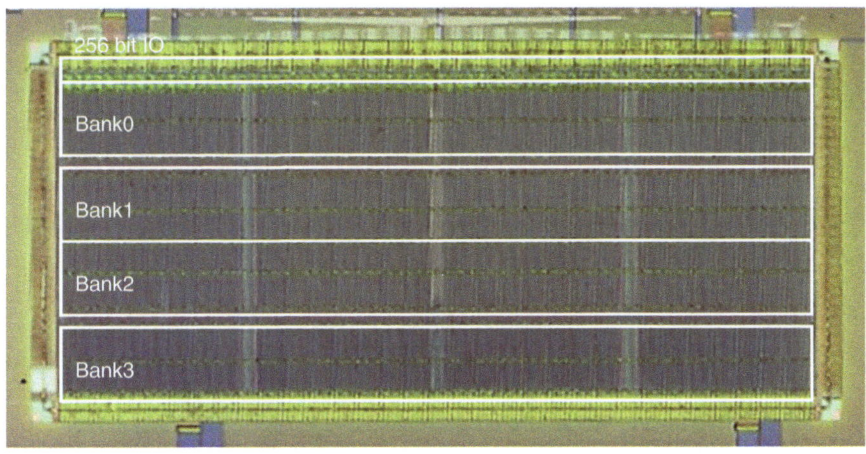

Process Technology	0.13-mm 6-metal 1-poly triple-well dual-Vth CMOS
Gate Length	0.10 mm
Power Supply	1.2 V
Vth	0.30 V / -0.30 V (low Vth nMOS / pMOS) 0.40 V /-0.40 V (high Vth nMOS / pMOS)
Memory Cell Size	$2.52 \times 0.84 = 2.12$ mm^2(lithographically symmetrical Cell)
Memory Cell Vth	0.40 V (nMOS) / -0.55 V (pMOS)
SRAM Macro Size	1 Mbit (1150×2830 mm^2)
Macro Structure	256 columns $\times$ 256 rows $\times$ 4 arrays $\times$ 4 banks

Fig. 4.14 Photograph and features of the prototype chip

low-voltage operation is described in this section. After the problems in conventional memory cells in low-voltage operation are described, a new technique for solving this problem, namely a lithographically symmetrical (LS) cell, is proposed. Results from our experimental evaluation of this technique are described.

4.3.1 Lithographically Symmetrical Memory Cell

This section describes the problems in conventional memory cells in low-voltage operation and our technique for solving this problem, namely a lithographically symmetrical cell. Figure 4.15 compares the LS-cell with conventional cell.

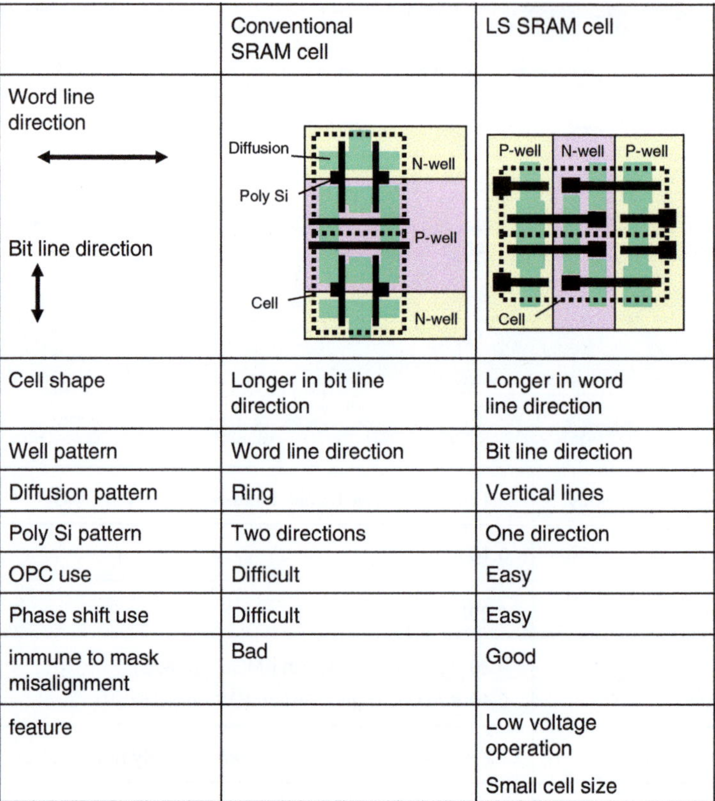

	Conventional SRAM cell	LS SRAM cell
Cell shape	Longer in bit line direction	Longer in word line direction
Well pattern	Word line direction	Bit line direction
Diffusion pattern	Ring	Vertical lines
Poly Si pattern	Two directions	One direction
OPC use	Difficult	Easy
Phase shift use	Difficult	Easy
immune to mask misalignment	Bad	Good
feature		Low voltage operation Small cell size

Fig. 4.15 Comparison between conventional and LS cells

The LS-cell has the lowest aspect ratio (0.38) of conventional SRAM cells (1.24 [4.14]). The diffusion layouts of conventional SRAM cells have bends. Each layer of the LS-cell is a point of symmetry, and the patterns of the diffusion layer are straight with no bends. The LS-cell has one-direction poly-silicon layer patterns although the conventional cell has two-direction patterns. The layouts of the poly-silicon layer and the diffusion layer are very simple; hence, it is easy to apply advanced lithography techniques such as optical proximity correction (OPC) or phase shift. Therefore, pattern fluctuation, which appears on silicon during photolithography, is reduced. As a result, the LS-cell is not influenced by photolithography misalignment and has good electrical balance. On the other hand, conventional cells have suffered from electrical imbalance. As a result, a new SRAM cell cannot provide a large enough noise margin at low voltage. Figure 4.16 shows SEM photographs of the LS-cell using a four-metal 0.18-μm enhanced CMOS technology [4.15]. The pitch from the first to the fourth metals is 0.52μm, and the memory cell size is just under 4.3 μm^2. Figure 4.16a shows its diffusion layer (vertical lines) and its poly-silicon layer (horizontal lines). Figure 4.16b shows the first metals (used for word lines)

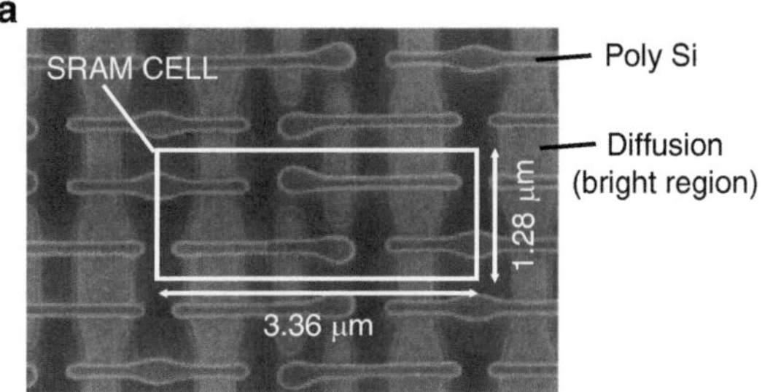

Fig. 4.16 SEM photograph of a lithographically symmetrical cell (LS-cell). Cell size is 4.30 μm^2; (**a**) diffusion and poly-Si layers, (**b**) first and second metals

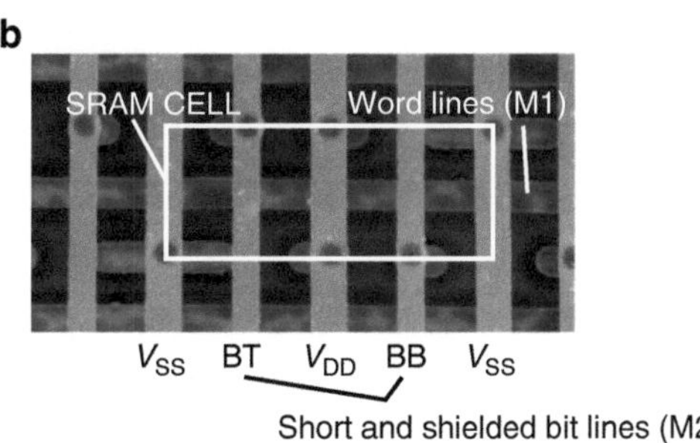

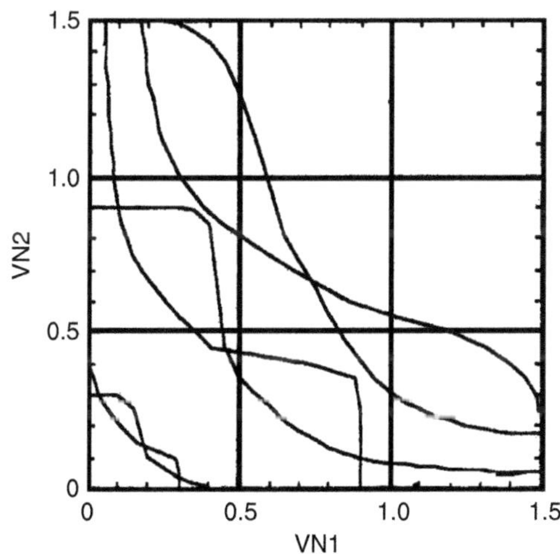

Fig. 4.17 Measured butterfly curves (typical)

and the second metals (used for bit lines and power supply). Figure 4.17 shows the measured butterfly curves of a typical LS-cell: even at 0.3 V, the memory cell has a large enough noise margin. The threshold voltage of the memory cells is 0.5 V.

There are many other advantages of the LS-cell. Its bit lines are shorter because of the low-aspect ratio; hence, parasitic capacitance of bit lines is reduced and read and write access is faster. Crosstalk between bit lines is drastically reduced because bit lines are shielded by V_{DD} lines and V_{SS} lines, and noises on the V_{DD}

Fig. 4.18 SEM photograph of diffusion and poly-Si layers of 45 nm SRAM cell. (**a**) Cell size 0.327 μm^2; (**b**) cell size 0.245 μm^2

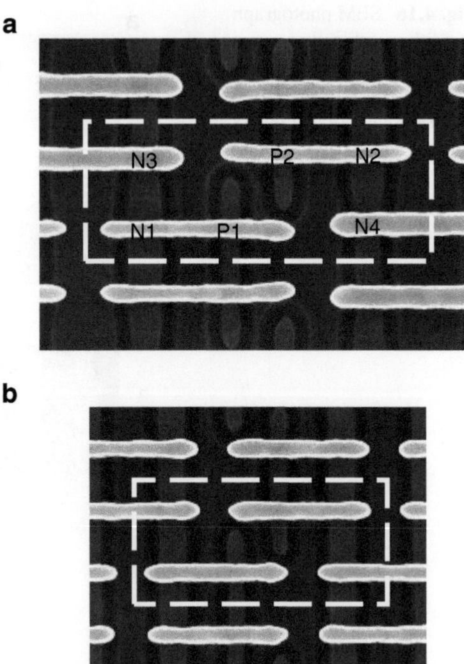

lines and V_{SS} lines are reduced because these lines run on orthogonal word lines and every memory cell's read current flows on each V_{DD} line and each V_{SS} line. However, its disadvantage is that the memory cell's word line becomes longer than that of a conventional SRAM cell. Even if this disadvantage is considered, the above-mentioned advantages lead to a reduction in the delay in total access time by 13%, compared to that of a conventional SRAM cell at 1.5 V. By these reasons, the LS-cell has been the common cell for the advanced CMOS processes since 90 nm node.

Figure 4.18 shows SEM photographs of SRAM cells using a 45-nm CMOS technology [4.16]. The cell sizes are (a) 0.327 μm^2 and (b) 0.245 μm^2. Each layer of the SRAM cell is a point of symmetry, and the patterns of the diffusion layer are straight with no bends. The SRAM cell has one-direction poly-silicon layer patterns. The layouts of the poly-silicon layer and the diffusion layer are very simple.

References

4.1. K.C. Chum, et al., A sub-0.9V logic-compatible embedded DRAM with boosted 3T gain cell, regulated bit-line write scheme and PVT-tracking read reference bias, in *Symposium on VLSI Technology Digest*, June 2009, pp. 134–135

4.2. S.-H. Lo, D.A. Buchanan, Y. Taur, W. Wang, Quantum-mechanical modeling of electron tunneling current from the inversion layer of ultra-thin-oxide nMOSFET's. IEEE Electron Device Lett. **18**(5), 209–211 (1997)

4.3. D.J. Frank, Device scaling, leakage currents, and joint technology and system optimization, in *2002 VLSI Circuits Symposium Short Course*, June 2002

4.4. M. Rosar, B. Leroy, G. Schweeger, A new model for the description of gate voltage and temperature dependence of gate-induced drain leakage (GIDL) in the low electric field region. IEEE Trans. Electron Dev. **47**(1), 154–159 (2000)

4.5. J. Maserjian, Tunneling in thin MOS structures. J. Vac. Sci. Technol. **11**(6), 996–1003 (1974)

4.6. J. Sune, P. Olivo, B. Ricco, Quantum-mechanical modeling of accumulation layers in MOS structure. IEEE Trans. Electron Dev. **39**(7), 1732–1739 (1992)

4.7. G.A.M. Hurkx, D.B.M. Klaassen, M.P.G. Knuvers, A new recombination model for device simulation including tunneling. IEEE Trans. Electron Dev. **39**(2), 331–338 (1992)

4.8. K. Nii, H. Makino, Y. Tujihashi, C. Morishima, Y. Hayakawa, H. Nunogami, T. Arakawa, H. Hamano, A low-power SRAM using auto-backgate-controlled MT-CMOS, in *Proceedings of the International Symposium on Low Power Electronics and Devices*, August 1998, pp. 293–298

4.9. H. Yamauchi, T. Iwata, H. Akamatsu, A. Matsuzawa, A 0.5 V single power supply operated high-speed boosted and offset-grounded data storage (BOGS) SRAM cell architecture. IEEE Trans. VLSI Syst. **5**(4), 377–387 (1997)

4.10. D.H. Kim, S.J. Kim, B.J. Hwang, S.H. Seo, J.H. Choi, H.S. Lee, W.S. Yang, M.S. Kim, K.H. Kwak, J.Y. Lee, J.Y. Joo, J.H. Kim, K. Koh, S.H. Park, J.I. Hong, Highly manufacturable 32Mb ULP-SRAM technology by using dual gate process for 1.5V Vcc operation, in *Symposium on VLSI Technology Digest*, June 2002, pp. 118–119

4.11. K. Osada, Y. Saitoh, E. Ibe, K. Ishibashi, 16.7-fA/cell tunnel-leakage-suppressed 16-Mbit SRAM for handling cosmic-ray-induced multi-errors. IEEE J. Solid-State Circuits **38**(11), 1952–1957 (2003)

4.12. M. Yamaoka, Y. Shinozaki, N. Maeda, Y. Shimazaki, K. Kato, S. Shimada, K. Yanagisawa, K. Osada, A 300-MHz 25 − μA/Mb-leakage on-chip SRAM module featuring process-variation immunity and low-leakage-active mode for mobile-phone application processor. IEEE J. Solid-State Circuits **40**(1), 186–194 (2005)

4.13. T.D. Burd, et al., A dynamic voltage scaled microprocessor system. IEEE J. Solid-State Circuits **35**, 1571–1580 (2000)

4.14. T. Uetake, et al., A 1.0ns access 770MHz 36Kb SRAM macro, in *Symposium on VLSI Circuits Digest*, June 1999, pp. 109–110

4.15. K. Osada, J.L. Shin, M. Khan, Y. Liou, K. Wang, K. Shoji, K. Kuroda, S. Ikeda, K. Ishibashi, Universal-Vdd 0.65–2.0-V 32-kB cache using voltage-adapted timing-generation scheme and a lithographical-symmetric cell. IEEE J. Solid-State Circuits **36**(11), 1738–1744 (2001)

4.16. K. Nii, M. Yabuuchi, Y. Tsukamoto, S. Ohbayashi, S. Imaoka, H. Makino, Y. Yamagami, S. Ishikura, T. Terano, T. Oashi, K. Hashimoto, A. Sebe, G. Okazaki, K. Satomi, H. Akamatsu, H. Shinohara, A 45-nm bulk CMOS embedded SRAM with improved immunity against process and temperature variations. IEEE J. Solid-State Circuits **42**(1), 180–191 (2008)

Chapter 5
Low-Power Array Design Techniques

Koji Nii, Masanao Yamaoka, and Kenichi Osada

Abstract Thischapter introduces circuit technologies that enhance electric stability of the cell, the latesttechnologies that provide moderate timing generation, as well as larger cell stability. In Sect. 5.1, the voltage-adapted timing-generation scheme with plural dummy cells for the wider-voltage-range operation is introduced. The effect of increasing the number of activated dummy cells on the dummy-bitline-driving-time fluctuation is described. Detailed circuit diagrams of the dummy-column cell and edge-column cell are shown. The cache was fabricated using $0.18 - \mu m$ enhanced CMOS technology. The cache chip can continuously operate from 0.65 to 2.0 V; its operating frequency and power are from 120 MHz at 1.7 mW and0.65 V to 1.04 GHz at 530 mW and 2.0 V.

In the next section, read and write stability assisting circuits are discussed. Local variation of transistors is one of the serious issues in sub 100 nm era. This section describes SRAM assist circuits that enhance stability of the cell despite the local variation. Source bias circuit is introduced to enhance the stability in write operation, while word line lowering enhances SNM in read operation. These circuit techniques do not need any additional supply voltages. A test chip is designed and fabricated using 45-nm CMOS technology. It is achieved over 100 mV improvement for the SNM and 35 mV for the write margin. Compared to the conventional

K. Nii (✉)
Renesas Electronics Corporation, 5-20-1, Josuihon-cho, Kodaira, Tokyo 187-8588, Japan
e-mail: koji.nii.uj@renesas.com

M. Yamaoka
1101 Kitchawan Road, Yorktown Heights, NY 10598, USA
e-mail: masanao.yamaoka@hal.hitachi.com

K. Osada
Measurement Systems Research Department, Central Research Laboratory, Hitachi, Ltd., 1-280, Higashi-koigakubo, Kokubunji-shi, Tokyo 180-8601, Japan
e-mail: kenichi.osada.aj@hitachi.com

K. Ishibashi and K. Osada (eds.), *Low Power and Reliable SRAM Memory Cell and Array Design*, Springer Series in Advanced Microelectronics 31, DOI 10.1007/978-3-642-19568-6_5, © Springer-Verlag Berlin Heidelberg 2011

assist circuit, the cell current at the worst case condition was improved by 83%. A more stable functionality of 512-kb SRAM macros with 0.245 and 0.327 μm^2 is observed. The minimum operating voltage in the process- and temperature-worst case condition is improved by 170 mV, and its variation to 60 mV, confirming a high immunity against process and temperature variations with less than 10% area overhead.

In Sect. 5.3, an array boost technique is demonstrated for low-voltage operation. The technique enhances stability of SRAM memory cells.

Section 5.4 introduces design techniques of dual-port SRAM cell array, which is often used in recent SOC as data memory for graphics engines. The electrical stability of the dual-port SRAM must be considered more seriously than the single port memory. Thereby, a new array design technique for a synchronous DP-SRAM was demonstrated. Using 65-nm CMOS technology, 32-kB DP-SRAM macros were designed and subsequently fabricated. This process yielded the smallest 8T DP-cell and the highest bit density ever reported in the 65-nm era. Test results show that the speed penalty was negligible; standby leakage was reduced by 27% because of the small cell size.

Acronyms

BOX	Buried OXide
CAD	Computer Aided Design
CADDETH	Computer Aided Device DEsign in THree dimensions
CHB	CHecker Board
CHBc	CHecker Board complement
CMOS	Complementary Metal Oxide Semiconductor
CORIMS	COsmic Radiation IMpact Simulation
DC	Direct Current
DRAM	Dynamic Random Access Memory
ECC	Error Checking and Correction
EFR	Electric-Field-Relaxation
FD-SOI	Fully Depleted Silicon On Insulator
FF	Flip Flop
FinFET	Fin Filed Effect Transistow
FPGA	Field Programmable Gate Array
GIDL	Gate-Induced Drain Leakage
INC	Intra-Nuclear Cascade
LER	Line Edge Roughness
MBU	Multi-Bit Upset
MCBI	Multi-Coupled Bipolar Interaction
MCU	Multi-Cell Upset

MNT	Multi Node Transient
MOSFET	Metal Oxide Silicon Field Effect Transistor
NMOS	Negative channel Metal Oxide Silicon
OPC	Optical Proximity Correction
PCSE	Power Cycle Soft-Error
PMOS	Positive channel Metal Oxide Silicon
RDF	Random Dopant Fluctuation
RBL	Read Bit Line
RWL	Read Word Line
S.A.	Sense Amplifier
SBU	Single Bit Upset
SoC	System-on-Chip
SRAM	Static Random Access Memory
SEB	Single Event Burnou
SECIS	SElf-Consistent Integrated System for neutron-induced soft-error
SEE	Single Event Effect
SEFI	Single Event Functional Interruption
SEGR	Single Event Gate Rupture
SEL	Single Event Latchup
SER	Soft-Error Rate
SET	Single Event Transient
SEU	Single Event Upset
SESB	Single Event Snap-Back
SNM	Static Noise Margin
SOI	Silicon on Insulator
STI	Shallow Trench Isolation
TMR	Triple Module Redundancy
ULSI	Ultra-Large Scale Integration
Vth	Threshold Voltage
W.A.	Write Amplifier
WL	Word Line

5.1 Dummy Cell Design

5.1.1 Problem with Wide-Voltage Operation

As power supply voltage is reduced, the threshold voltage of a MOS transistor
should also be reduced for high-speed operation. However, the threshold voltage
of memory cells must be kept at a minimum of 0.5 V to reduce the leakage current

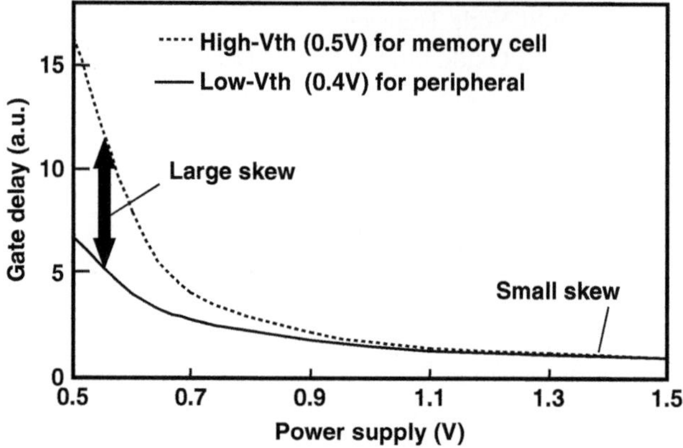

Fig. 5.1 Gate delay as a function of power supply voltage

and to keep adequate static noise margins. On the other hand, to attain high speed at low voltage, the threshold voltage of peripheral circuits can be reduced to 0.4 V. Figure 5.1 shows the gate delay of inverter circuits as a function of power supply voltage. The dotted line represents the delay due to the high-threshold-voltage (0.5 V) transistor used for the memory cell. The solid line represents the delay due to the low-threshold-voltage (0.4 V) transistor used for peripheral circuits. At a power supply voltage of 1.5 V, the gate delay of the low-threshold-voltage transistor is almost the same as that of the high-threshold-voltage transistor. At 0.5 V, however, there is a large skew between the delays.

This skew could cause activation failure of a sense amplifier during wide-range-voltage operation. To avoid such failure, a voltage-adapted timing-generation scheme using plural dummy cells was developed. The point of this scheme is that the delay driven by high-threshold voltage in the data read path is the same as that driven by high-threshold voltage in the control path. Thus, it can compensate the skew between the data read path and the control path under wide-range voltage.

5.1.2 *Block Diagram and Operation of Voltage-Adapted Timing-Generation Scheme*

Figure 5.2 shows a block diagram of a half side of one bank of the cache data array, which is composed of four banks. Each bank is composed of 256 word lines by 256 bit columns. When the clock signal (CK) changes from "L" to "H," the signal (dec_en) is generated in a D-flip flop (D-FF) and it activates predecoders and selects a word line (wl). Signal voltages then are produced on bitline pairs (bt, bb) by the selected memory cells (MCs) in the regular SRAM array. All these operations make up the data read path.

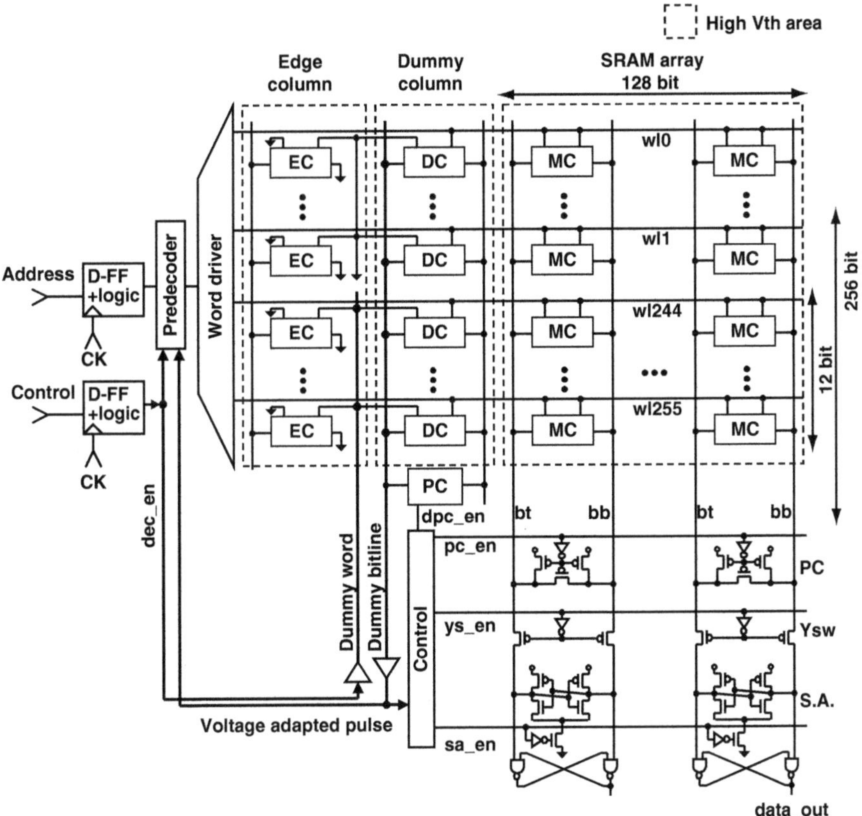

Fig. 5.2 Block diagram of half side of a bank

In the control path, the signal (dec_en) also activates the dummy word line, which runs parallel to the bitline (bt, bb). The dummy word line activates 12 dummy cells (DCs) on the dummy column. The activated DCs drive the dummy bitline, whose capacitance is identical to the regular bitline (bt, bb). The dummy bitline then becomes the voltage-adapted pulse. The voltage-adapted pulse is used for the sense-amplifier-enable signal (sa_en), the pre-charge-reset signal (pc_en), and the word line-reset signal. The most important point is that the delay due to the high-threshold-voltage memory cells (DCs) is also included in the control path that activates the sense amplifier.

Detailed circuit diagrams of the dummy-column cell (DC) and edge-column cell (EC) are shown in Fig. 5.3. The dummy column is used as the electrical dummy column, while the edge column is used as the optical dummy. As a result of this circuit scheme, the layouts of the diffusion layer and polysilicon layer in the regular SRAM array, the dummy column, and the edge column can be kept regular. Therefore, the dummy-cell current is identical to the memory cell current in the regular SRAM array. In these circuits, the dummy word line runs vertically to

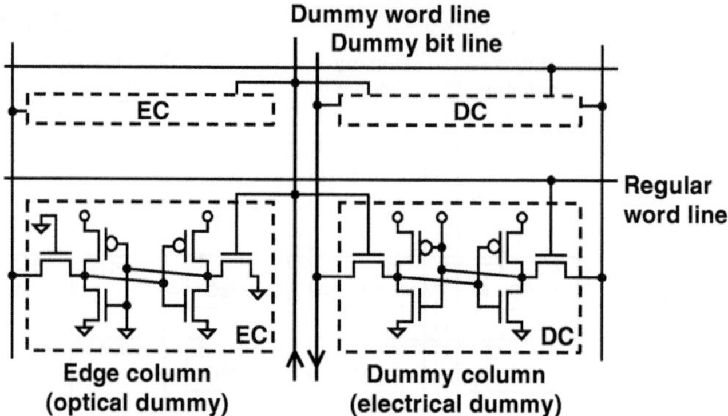

Fig. 5.3 Diagrams of dummy and edge column circuit

activate the plural dummy cells. The pattern of the bit line in the edge column is used as the dummy word line, which is different from a regular word line; however, this difference does not cause a problem because the word line is driven by the same-threshold-voltage transistors as those of the peripheral circuits.

5.1.3 Timing Diagram and Effect of Voltage-Adapted Timing-Generation Scheme

Figure 5.4 compares the timing diagrams of the data array operation at power supply voltages of 2.0 and 0.65 V. At 2.0 V, bitline-driving time is 24% of the total access time. However, bitline-driving time increases to 48% at 0.65 V because of the memory cell with a high-V_{th} transistor. In our scheme, high-V_{th} dummy cells drive one part of the timing-generation path; so the sense amplifier is activated suitably even at low voltage.

Although a dummy-cell structure has already been developed [5.1], a single dummy cell suffers from cell current fluctuation as shown in Fig. 5.5. Such dummy-cell current fluctuation results in the fluctuation of the control timing generated by the dummy cell. Figure 5.6 shows the effect of increasing the number of activated dummy cells on the dummy-bitline-driving-time fluctuation (normalized by total access) at 1.5 V. The fluctuations of memory cell current were calculated using a standard deviation of 4%, which was obtained from the distribution shown in Fig. 5.5. The dummy-bitline-driving-time fluctuation is reduced from 17.5% to 5% when the number of dummy cells is increased from 1 to 12, which is used in this work because the fluctuation of dummy-cell current is averaged. This fluctuation reduction can reduce the timing margin for activating the sense amplifier and reduce the delay by 12.5% compared to that of conventional circuits with a single dummy cell [5.1]. It is thus concluded from Figs. 5.4 and 5.6 that the influences of threshold-voltage difference and the variation in dummy-cell current on circuit performance can be reduced by our new scheme.

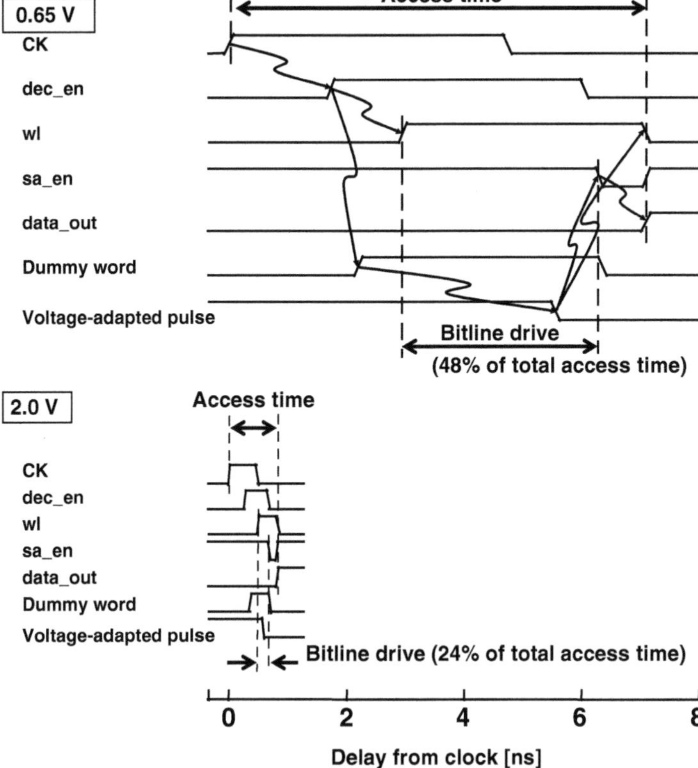

Fig. 5.4 Timing diagrams of the cache operation at 2.0 V and 0.65 V

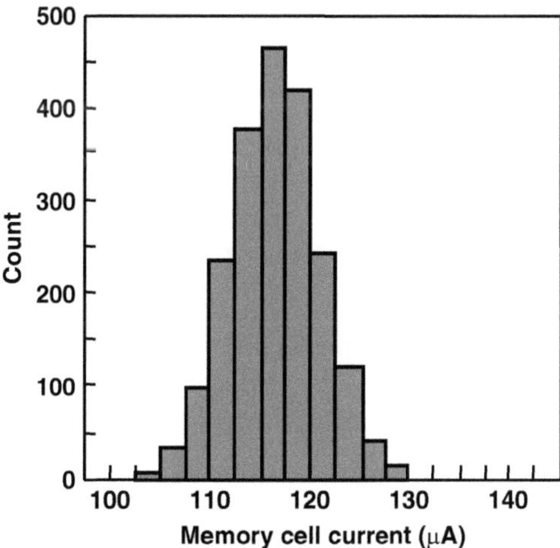

Fig. 5.5 Measured SRAM cell current distribution

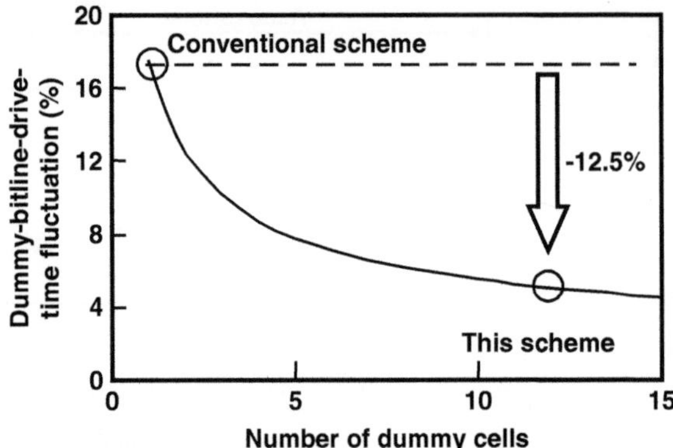

Fig. 5.6 Effect of plural dummy cells

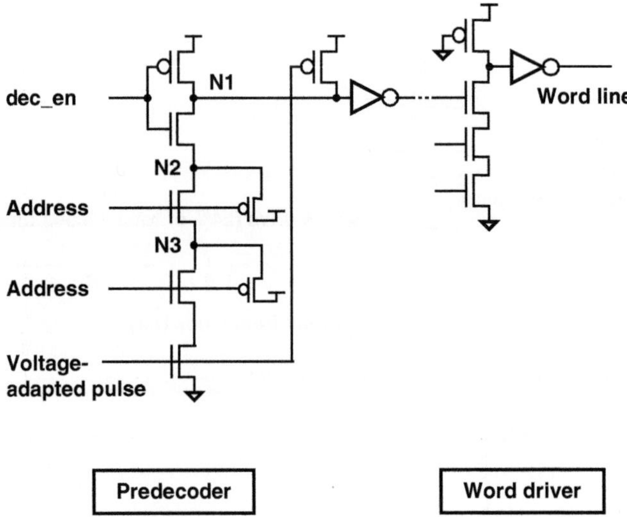

Fig. 5.7 Predecoder and word driver circuits

5.1.4 Predecoder and Word-Driver Circuits

Figure 5.7 schematically shows the predecoder circuits and the word-driver circuit. Word line reset is carried out only in the predecoder; so the area for the reset circuit is minimized compared to that in which a word line is reset in the word driver. The word lines are switched off by the voltage-adapted pulse just after the sense amplifier is activated. Thus, because the voltage-adapted pulse is used for reset, the word line pulse is minimized according to operation voltage, and the wasted power in the memory cells can be minimized in spite of a wide range of operating voltages.

For reset, four NMOS transistors are connected in series. Because only two PMOS transistors are connected at N1 instead of four PMOS transistors, the delay of N1 discharge, as well as the delay of word activation, is decreased. For low-voltage operation, small PMOS transistors are connected at N2 and N3 to avoid charge-shearing problems.

5.1.5 Results

The developed 32-kB cache was fabricated using a 4-metal $0.18 - \mu$m enhanced CMOS technology. The gate lengths of the nMOS and pMOS devices are both $0.14\,\mu$m. The threshold voltage of the memory cells is 0.5 V, whereas the threshold voltage of the peripheral transistors is 0.4 V. The pitch from the first to the fourth metals is $0.52\,\mu$m, and the memory cell size is just under $4.3\,\mu\text{m}^2$. Figure 5.8 is a photograph of the fabricated cache chip. The cache includes a data array and a tag array. The data array is composed of four 8 kB banks, each bank is composed of one

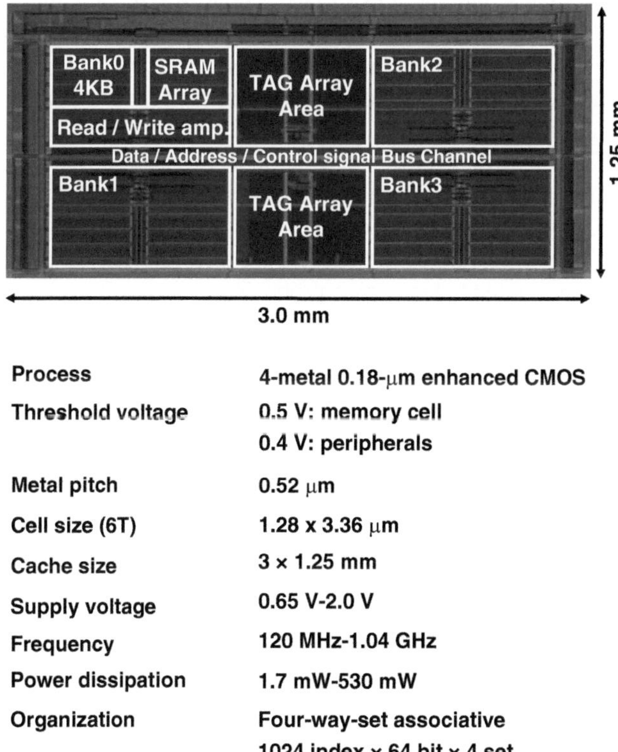

Process	4-metal 0.18-μm enhanced CMOS
Threshold voltage	0.5 V: memory cell
	0.4 V: peripherals
Metal pitch	0.52 μm
Cell size (6T)	1.28 x 3.36 μm
Cache size	3 × 1.25 mm
Supply voltage	0.65 V-2.0 V
Frequency	120 MHz-1.04 GHz
Power dissipation	1.7 mW-530 mW
Organization	Four-way-set associative
	1024 index × 64 bit × 4 set

Fig. 5.8 Photograph and specifications of developed cache

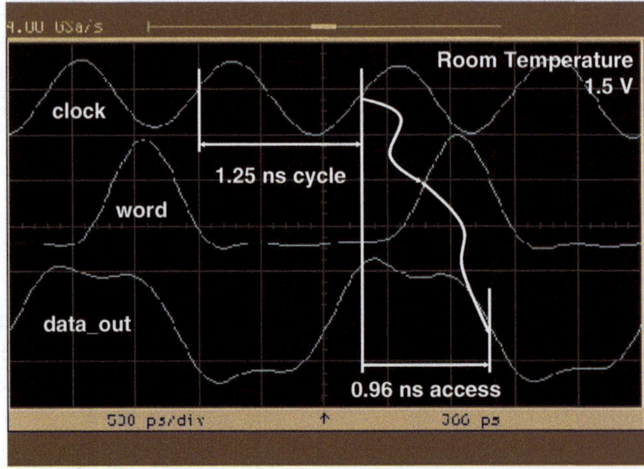

Fig. 5.9 Waveforms of read operation

double word, the bank is composed of 256 word lines by 256 bit columns, the cache area is 3×1.25 mm, and well taps are placed every 64 rows in each memory array.

The data array has an architecture that minimizes power consumption. Only one bank is decoded by the two index address bits that are available before the cache access, and the other three banks are disabled. This approach saves power consumption by about 30% over the conventional approach, which decodes all the banks. The main power saving is gained from disabling array-control signals and the bit-line swings of unselected banks.

Read-operation waveforms of the cache data array are shown in Fig. 5.9. The waveforms were measured using a pico-probe. Read access from the clock to the output of a sense amplifier (data_out) is 960 ps at 800 MHz and 1.5 V. An on-chip PLL was used to obtain the shmoo plot of the data array (Fig. 5.10). The cache chip operates from 120 MHz at 0.65 V to 1.04 GHz at 2.0 V. Figure 5.11 shows the data array access time as a function of supply voltage under varying substrate voltage (mVbb) of the NMOS transistor in the memory cell. When mVbb is negative, the threshold voltage increases and memory cell current is reduced; as a result, access time is increased. When mVbb is positive, the threshold voltage decreases and memory cell current is increased; as a result, access time is decreased. This figure shows that our voltage-adapted timing-generation scheme successfully generates the control timing according to memory cell current.

Figure 5.12 shows the measured power dissipation of the data array without bus drive at the minimum voltage for each frequency. Moreover, conventional power dissipation was simulated at a constant voltage of 2.0 V for each frequency. Clearly, measured power dissipation at 120 MHz and 0.65 V is 1.7 mW and that at 1.04 GHz and 2.0 V is 530 mW. At 120 MHz frequency, measured power dissipation is reduced by 97% compared with conventional power dissipation.

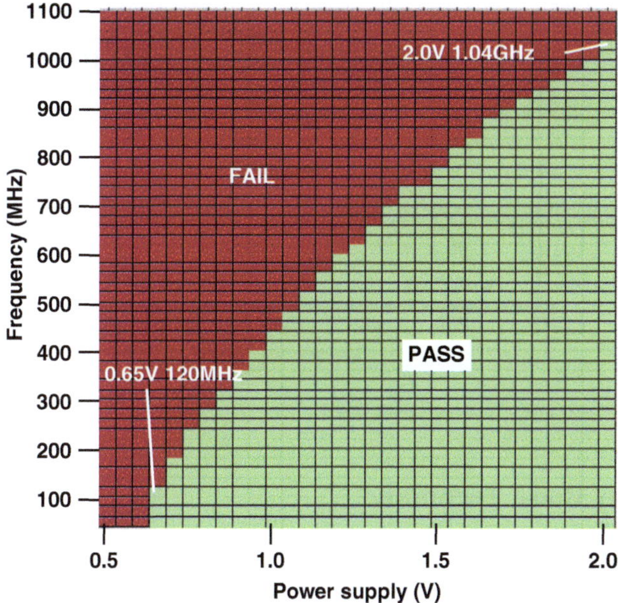

Fig. 5.10 Shmoo plot

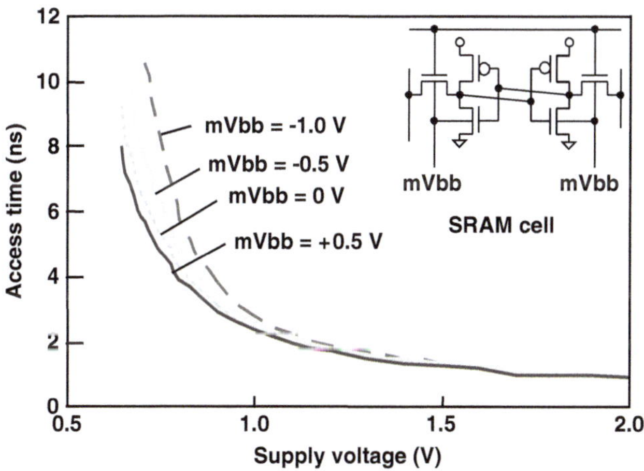

Fig. 5.11 Access time versus supply voltage

Figure 5.8 lists the specifications of the developed cache. The cache size is 3.75 mm^2. It operates from 120 MHz at 1.7 mW and 0.65 V to 1.04 GHz at 530 mW and 2.0 V. The organization of the cache is four-way-set associative and 1,024 indexes.

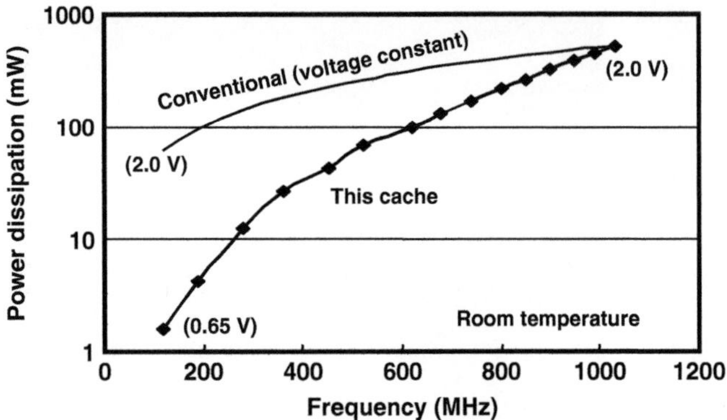

Fig. 5.12 Measured power dissipation of cache

5.2   Array Boost Technique

In low-voltage SoCs, the V_{th} must be reduced to maintain operation speed. In consequence, the SNM of SRAM memory cell is deteriorated by the low supply voltage and the low V_{th}. Butterfly curves using two types of supply voltage are shown in Fig. 5.13a. The width of squares within the curves shows the SNM [5.2]. The solid lines show the butterfly curves with 1.0-V supply voltage, and the dashed lines show the butterfly curves with 0.5-V supply voltage. The low supply voltage narrows the square within curves, and it decreases the SNM. Figure 5.13b shows the butterfly curves using lower V_{th} MOSFET, solid curves, and higher V_{th} MOSFET, dashed curves. The low V_{th} MOSFET crushes the space between two lines, and it decreases the SNM. To improve the SNM, array boost technique is used. Figure 5.14 shows the circuit diagram of the array boost technique, where V_a is the power supply voltage of SRAM cell array and V_{dd} is the power supply voltage of peripheral circuits, which are used to access the SRAM cells.

In the array boost technique, V_{dd} is supplied to a boosted voltage generator, and the generator boosts V_{dd} to $V_a(= V_{dd} + \Delta V_a)$ and supplies V_a to the SRAM cell power supply (the source line of the load MOS of the memory cell). The supply voltage of the SRAM array is always boosted whenever the SRAM circuit is activated. Since the boosted voltage (ΔV_a) is a small and fixed value such as 0.1V, it does not affect reliability of the MOSFETs. The memory cell can thus be composed of small MOSFETs that can endure only low voltage. This keeps the scaling of fabrication process and enables the memory cell area to be decreased with the scaling of logic circuit. Figure 5.15 shows the relationship between V_{dd} and V_a. The electrical β-ratio means the ratio of conductance of driver MOS to that of access MOS. The details about the electrical β-ratio are explained later in this section.

Change of butterfly curves when implementing the array boost technique is shown in Fig. 5.16. The butterfly curves are not symmetrical because they consider

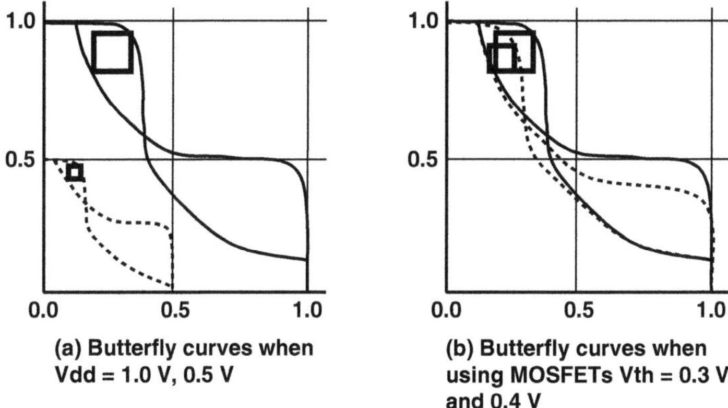

Fig. 5.13 Butterfly curves

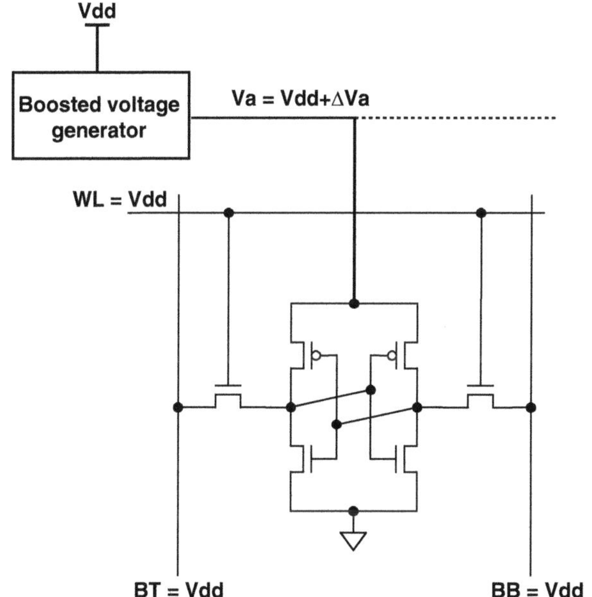

Fig. 5.14 Array boost technique

$3\text{-}\sigma V_{th}$ variation in one memory cell. The curves, node1va and node2va, are the voltage of one storage node when changing the voltage of the opposite node in a memory cell using array boost technique. The curves, node1vdd and node2vdd, are the voltage without array boost technique. The SNM is calculated as the diagonally length of squares that is placed between two curves. The curves, node1va and node1vdd, are almost same around the square, so that the difference in the side

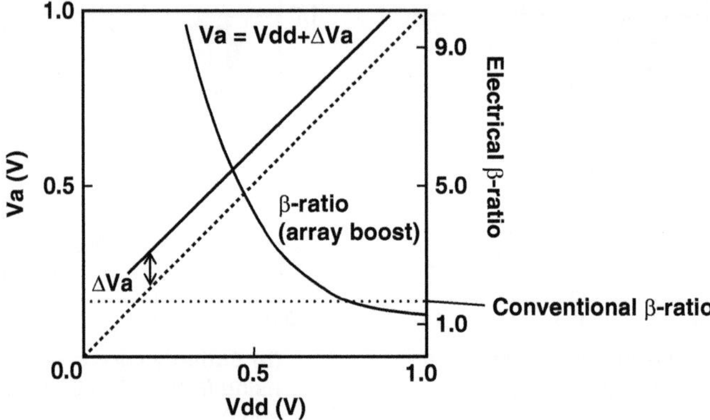

Fig. 5.15 Boosted array voltage and electrical β-ratio

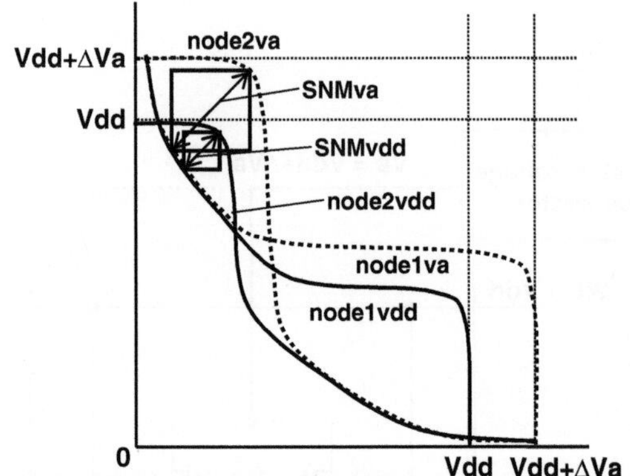

Fig. 5.16 Butterfly curves wi/wo array boost technique

length between two squares is ΔV_a. Therefore, SNMva (the SNM using the array boost technique) is calculated as:

$$\mathrm{SNMva} \approx \mathrm{SNMvdd} + \Delta V_\mathrm{a}\sqrt{2}. \tag{5.1}$$

For example, when $\Delta V_\mathrm{a} = 0.1$ V, the SNMva is calculated as:

$$\mathrm{SNMva} \approx \mathrm{SNMvdd} + 0.1\,\mathrm{V}\,\sqrt{2}, \tag{5.2}$$

$$= \mathrm{SNMvdd} + 0.07\,\mathrm{V}, \tag{5.3}$$

and SNM is improved by 70 mV. Using the array boost technique, the source to gate voltage (V_gs) of the driver MOS is higher than the V_gs of the access MOS, and the conductance of the driver MOS becomes larger than that of the access MOS. This conductance ratio is called as "electrical β-ratio." The increased electrical β-ratio

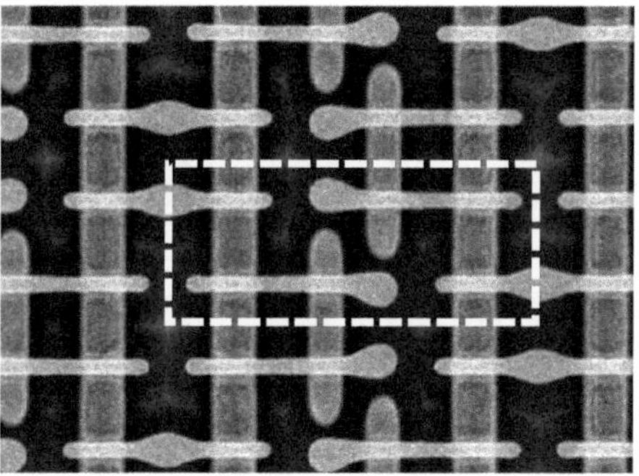

Fig. 5.17 SEM photograph of RD cell

makes SNM high. When the boosted voltage keeps a constant value, the electrical β-ratio is greatly increased in low supply voltage region. Figure 5.15 shows the electrical β-ratio and the dashed horizontal line shows the β-ratio of a conventional memory cell. In the region where V_{dd} is under 0.8 V, the electrical β-ratio becomes larger than conventional one. The large β-ratio in the low-voltage region provides high cell stability.

Conventionally, larger-than-1.0 β-ratio is realized by wider gate width of the driver MOS than that of the access MOS. Therefore, the diffusion layer in a memory cell has bending shape and it causes a transistor performance variation. When using the array boost techniques, the larger-than-1.0 β-ratio is realized electrically, and therefore, the difference between the gate width of the driver MOS and the access MOS is not necessary. Figure 5.17 shows an SEM photograph of rectangular diffusion (RD) cell. The diffusion layer has no bent and is in rectangular shape. The straight shape contributes to reduce the performance variation of the driver MOS and the access MOS.

Figure 5.18 shows the change in simulated SNM. The dark hatched stripe (1) shows the SNM of an RD-cell without an array boost technique. The breadth of this area is caused by 3-σV_{th} variation of the MOSFETs in an SRAM memory cell. In the region for which the supply voltage is less than 0.6 V, the SNM reaches almost 0 V. The light hatched stripe (2) in Fig. 5.18 shows the SNM with the array boost technique, in which the boosted voltage applied to SRAM array is 0.1 V higher than the supply voltage of the peripheral circuits. The SNM increases to almost 70 mV and this coincides with (5.3). The improvement of SNM is enough to operate this memory cell at 0.4-V supply voltage. Thus, the array boost technique improves the SNM of the SRAM cell, thereby realizing low voltage operation even when using the small β-ratio memory cell. In this way, the array boost technique can compensate for the SNM, which is deteriorated by increasing V_{th} variation, 1.0 β-ratio RD-cell and lower V_{th} MOSFET. Conventionally, we could not decrease the SRAM MOSFET's V_{th}, because the low V_{th} MOS deteriorates the SNM, and therefore, SRAM operating speed becomes lower than that of the logic circuits in the low-voltage region.

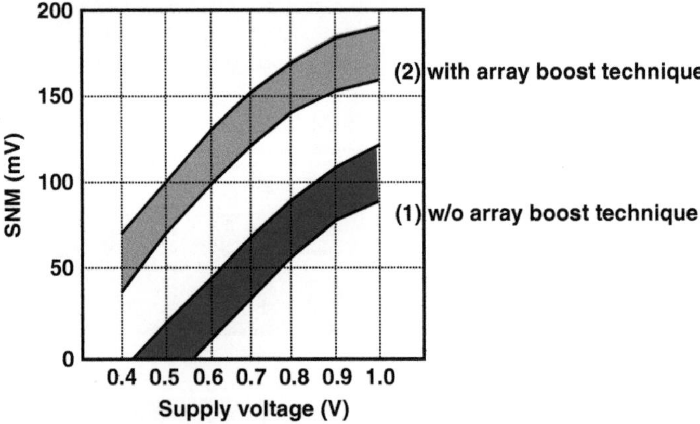

Fig. 5.18 SNM wi/wo array boost technique

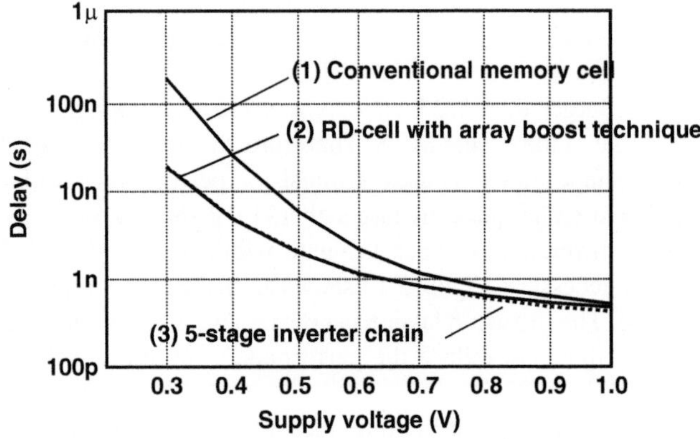

Fig. 5.19 Operating speed

Figure 5.19 shows the operating speed of SRAM and logic circuit. The solid line (1) shows the time for driving a bitline for 10% swing with a memory cell using a higher V_{th} MOSFET, and the dashed line (2) shows the same time with a memory cell using a lower V_{th} MOSFET and the dashed line (3) shows the delay of a five-stages inverter chain whose MOSFET's V_{th} is the same as the lower V_{th} SRAM. At V_{dd} of around 1.0 V, the speed of the two memory cells is nearly identical, but at V_{dd} of around 0.5 V, the speed of the memory cell using a higher V_{th} MOS becomes much slower than those of the RD-cell and the inverter chain. As shown in the figure, using the array boost technique, the memory cell MOSFET V_{th} to be reduced, thus increasing the SRAM speed to the same level as that of the logic circuits. Formerly, there were some techniques in which the voltage of SRAM array power supply was boosted (e.g. [5.3]).

Fig. 5.20 Prototype chip photograph and features

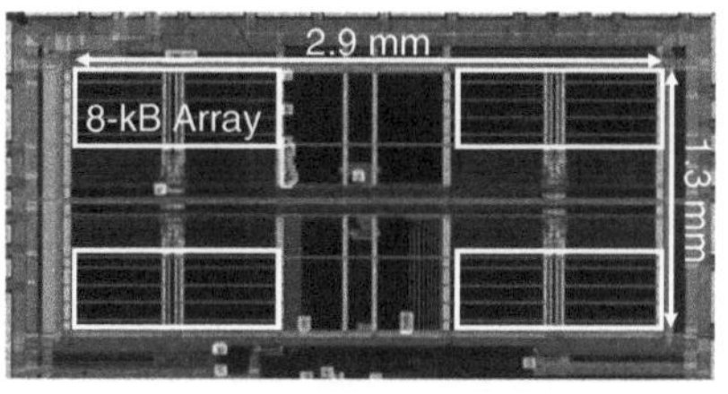

Process	6-metal 0.18-μm CMOS single poly
Vth	0.5 V (High Vth) / 0.4 V (Low Vth)
Memory cell Vth	0.4 V (nMOS) / 0.5 V (pMOS)

This array boost technique is based on a new concept that differs from those of the conventional array-boosting techniques. The concept is that the boosted voltage is so small that it does not affect memory cell MOSFET reliability. In the array boost technique, the boosted voltage (ΔV_a) is on a small level, e.g., 0.1 V, and therefore the memory cell MOSFET can be the same as that of the core logic circuits. If the V_a is much higher than V_{dd}, the MOSFET has to endure the higher voltage, and the MOSFET may be different from the MOSFET of the logic circuit. This compatibility decreases the cost of manufacturing LSIs and makes it possible to use the SRAM in the SoCs, and it keeps scaling of the memory cell, and therefore the memory cell can keep the smallest size in every fabricating process.

Figure 5.20 shows a photograph of 180-nm prototype chip with array boost technique.

5.3 Read and Write Stability Assisting Circuits

5.3.1 Concept of Improving Read Stability

The SRAM read margin is well expressed by the SNM. Figure 5.21 shows the simulated butterfly curves without local V_{th} variations using 45-nm node SPICE model at the two different conditions: one is of the fast-slow (FS) global process corner at high temperature (125°C), and the other is of slow–fast (SF) corner at low temperature (−40°C). Note that the FS (SF) means the combination of fast (slow) NMOS and slow (fast) PMOS. The supply voltage is 1.0 V minimum condition, the butterfly curves without local V_{th} variations show a good symmetry in bilateral direction, and the SNM strongly depends on the temperature and process global variations. In particular, the SNM has a minimum value at the FS corner, so that we hereafter deal with the SNM in the FS process condition. Figure 5.22 shows the SNM dependence on the NMOS V_{th} variation for three temperatures (−40, 27, 125°C). The SNM degrades at lower NMOS V_{th} or at higher temperature condition, while it improves at higher NMOS V_{th} or lower temperature condition.

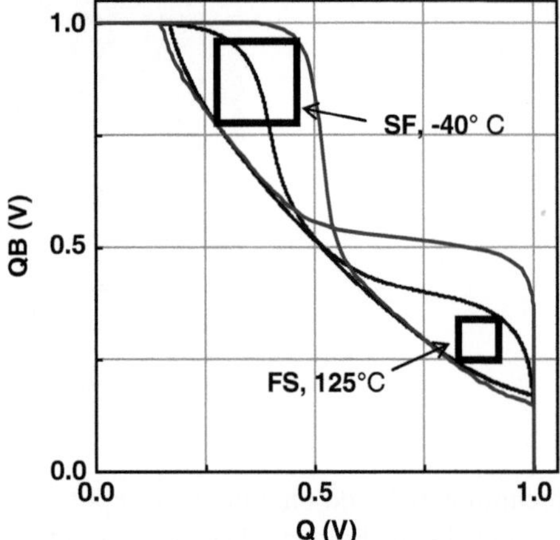

Fig. 5.21 6T-SRAM butterfly curves under worst and best conditions

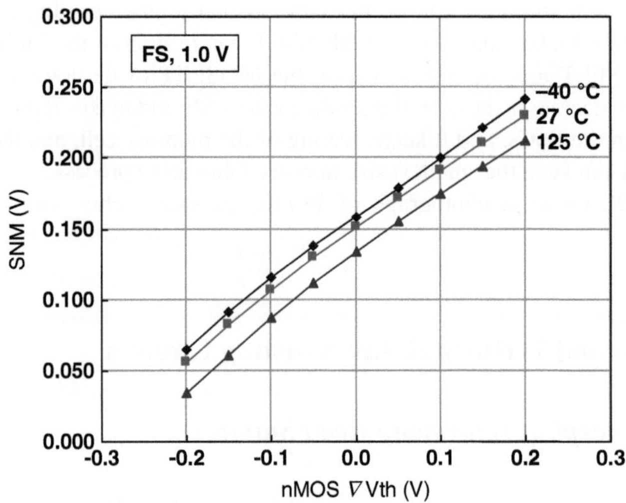

Fig. 5.22 Simulated static noise margin (SNM) versus threshold voltage of nMOS

In this way, by taking temperature and process global variation into consideration, we have to design the SRAM transistor size so that the operating margin becomes larger. As a result, the SNM should be ensured at the worst case of FS process condition at high temperature. Although we only discussed the read margin here, it is also necessary to examine the write margin precisely against temperature and process variation. By performing the similar analysis, one can obtain that the SF process corner at low temperature provides the worst condition of the write operation.

Next, we discuss a technique to enlarge the SNM at the presence of local V_{th} variation. In this section, we suggest the methodology to lower the voltage level of

Fig. 5.23 SNM improvement
by lowering word line voltage
(V_{WL}) in case of with/without
local V_{th} variations

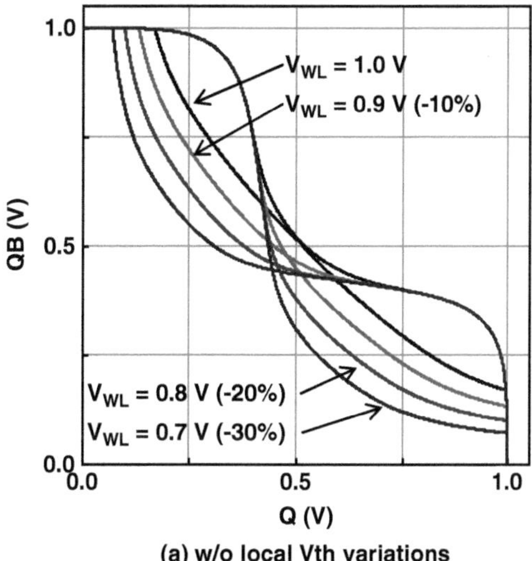

(a) w/o local Vth variations

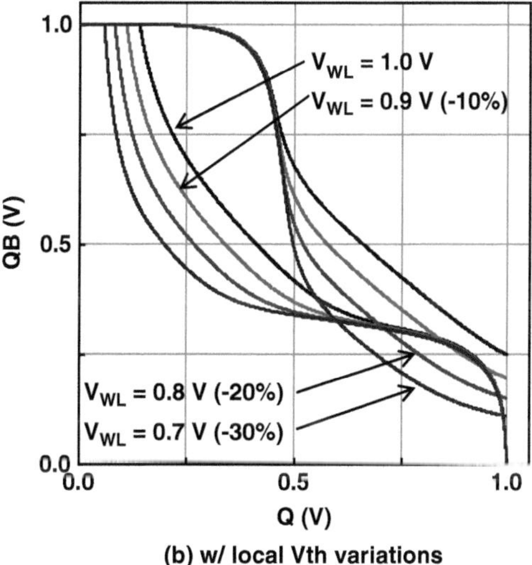

(b) w/ local Vth variations

the WL compared to that of the power line of flip-flop in an SRAM cell. Figure 5.23a
shows SNMs without local V_{th} variations at the worst condition (discussed above),
where we introduced the lowered voltage level to the WL. As is shown in this
graph, it is confirmed that the SNM is improved by lowering the WL level with
keeping the symmetric margin. In contrast, Fig. 5.23b indicates the result in the
same simulated condition, but the local V_{th} variation is additionally introduced
to each SRAM transistor. Because randomness of the local V_{th} variation destroys
the symmetry of the butterfly curve, the SNM without lowering the WL level in

Fig. 5.23a is reduced to be less than zero, indicating no expectation for SRAM read margin. Assuming that the randomness of the local V_{th} variation obeys the normal probability distribution, this deterioration of the SNM could be seen statistically if the total capacity of the SRAM memory array increases. In other words, the unit cell with asymmetric degradation of the SNM in 1-Mb SRAM array is more likely to appear than that in 1-Kb array [5.4]. Therefore, we need to reduce the WL level so that the SNM with local V_{th} variation becomes larger than zero. Using the results obtained in Fig. 5.23b, the WL level should be lowered by more than 20% compared to the supply voltage of 1.0 V.

Lowering the WL voltage accompanies with, at the same time, the drawbacks not only of the operation speed (or cell current) but also of the write operation. This is because the lowered WL reduces the ability of the access (pass-gate) transistor in an SRAM cell, which affects directly to the operation speed and write margin. Therefore, it is necessary to search for the optimum condition to enhance the SNM as large as possible, with a minimum loss of both the write margin and the operation speed. In the next section, we describe how to control the WL voltage level.

5.3.2 *Variation Tolerant Read Assist Circuits*

Figure 5.24 shows the read assist circuit (RAC) to control the word line voltage (V_{WL}). In the simple circuitry [5.5] as shown in Fig. 5.24a, the V_{WL} is lowered by plural pull-down NMOSs (called replica access transistors, RATs). Figure 5.25 represents the simulated V_{WL} dependence on the NMOS global V_{th} variations at high and low temperature, which provides two serious problems in the conventional circuit (see dotted lines). The first is a considerable decrease of V_{WL} in the FS condition at low temperature. An excessive lowering of the V_{WL} strongly enhances the SNM, which in turn degrades the write margin because of a low gate-overdrive of the access transistor. The second is also a decrease in the V_{WL} in the SS condition. In Fig. 5.25, the V_{WL} at SS, which is the worst condition of the cell current, becomes lower than that at TT due to its higher V_{thp}, resulting in a further degradation of the cell current.

We examine the reason why this unexpected V_{WL} degradation appears in the conventional circuit. Figure 5.26a is the simulated I-V curve for RAT and P0. Each simulation was performed at the worst 1.0 V condition for $-40°C$ and $125°C$. In the static condition, the current that flows through the P0 (IDD) is equivalent to that through the RAT (ISS), so that the V_{WL} is determined by the cross point between these two I-V curves. On the contrary to the RAT, the P0 has a strong dependence on the temperature, which causes a wide range of the V_{WL} fluctuation as appeared in Fig. 5.25a. In other words, the conventional circuits have disadvantage against not only process V_{th} variations but also the temperature fluctuation.

In order to overcome this problem, we propose a new circuitry shown in Fig. 5.24b. The RATs are introduced to the source of the WL driver, not to the

Fig. 5.24 Schematics of read assist circuits (RAC). (**a**) Original and (**b**) improved with passive resistance

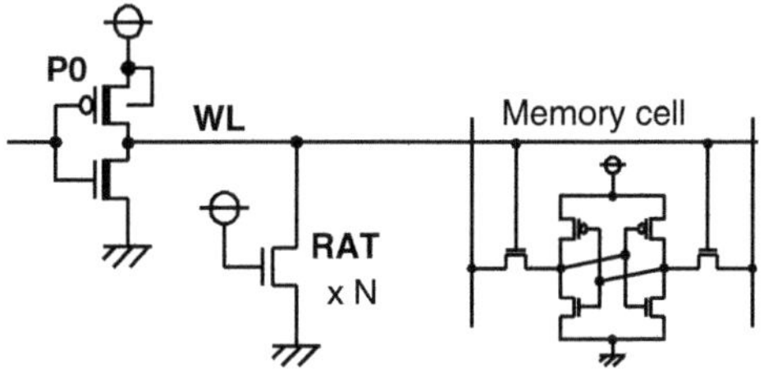

(a) Read assist circuit

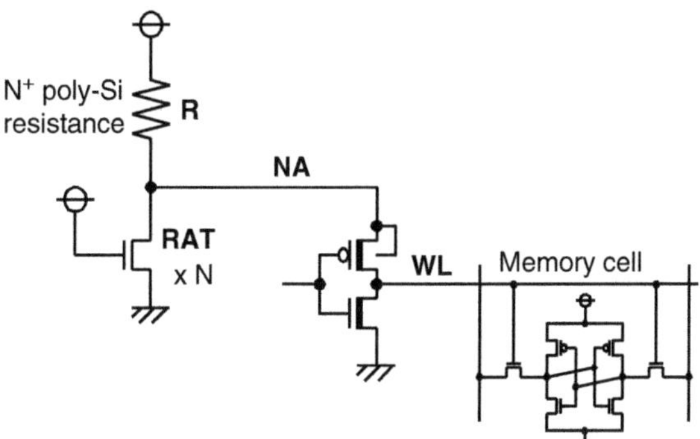

(b) Improved read assist circuit

WL in the memory cell array. In addition, we make use of the passive resistance elements (R) implemented by N type non-silicide polysilicon gate. This structure enables us to control the WL voltage reflecting both the process and the temperature variations. As shown in Fig. 5.26b, the IDD current obeys the ohmic characteristic, while the ISS current the transistor I-V curve. Due to the fact that the temperature dependence of the resistance element is much smaller than that of the drain current in the saturation region, we can suppress a V_{WL} fluctuation against temperature by eliminating the PMOS characteristic. Furthermore, it should be noted that the usage of the resistance elements has a great advantage on the V_{th} change which is responsible for the process variation. In general, the gate length of the polysilicon in the SS condition gets longer than that in the TT condition, which induces a higher V_{th} value. At the same time, the amount of the resistance element using polysilicon gate is expected to get lower because of the enlargement of cross-sectional area. From

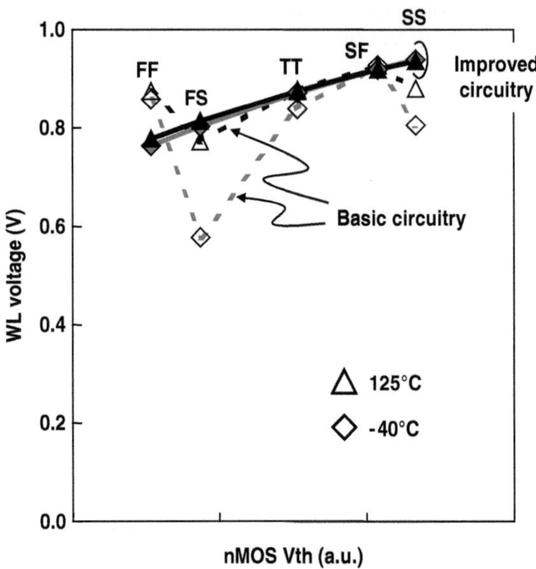

Fig. 5.25 Simulation result of WL voltage (V_{WL}) depending on nMOS V_{th}

Fig. 5.26b, it is easily obtained that a lower resistivity of R and a higher NMOS V_{th} (smaller ISS) at the SS corner lead to a higher V_{WL}, while a higher resistivity of R and a lower NMOS V_{th} at the FF corner lead to a lower V_{WL}. Combining these facts, we can conclude that the V_{th} of RATs dominates the V_{WL} value with a little temperature dependence. We have verified the validity of these V_{WL} behaviors through our simulation as shown in the Fig. 5.25 (see solid lines).

Figure 5.27 shows the overall architecture that implemented the proposed RAC. Additional circuitry denoted as gate controller (GC) has been introduced, which consists of serial resistors (R1 and R2) and the switching transistors (P1 and N1). By applying this circuitry, one can realize highly tolerant RAC against the PVT variations rather than a simple pull-down transistor.

Before explaining the functionality of the GC, we have to mention the layout of these critical resistors. Figure 5.28 represents the schematic layout view of our resistance elements. In order to reflect the process variation of the gate length L and width W of the access transistor in the SRAM memory cell, the line width of non-silicide polysilicon (R0, R2) and the non-silicide diffusion elements (R1) have the same size as L and W, respectively. In addition, the poly-Si pitch of R0 and R2 is equal to that of the SRAM memory cell, leading to the same resistance sensitivity in conjunction with the access transistor of the SRAM cell. Figure 5.29 indicates the resistance sensitivities of R0, R1, and R2 to the critical dimension (CD) shift of the gate and the diffusion sizes. The CD shift, which is caused by manufacturing variations in the lithography or etching process steps, influences the global V_{th} variations. For simplicity, we focus on the SS corner to explain the behavior of the additional GC. The SS corner means a high V_{th} and a low Ids condition, implying that the line width of the polysilicon gets larger and that of the diffusion gets smaller.

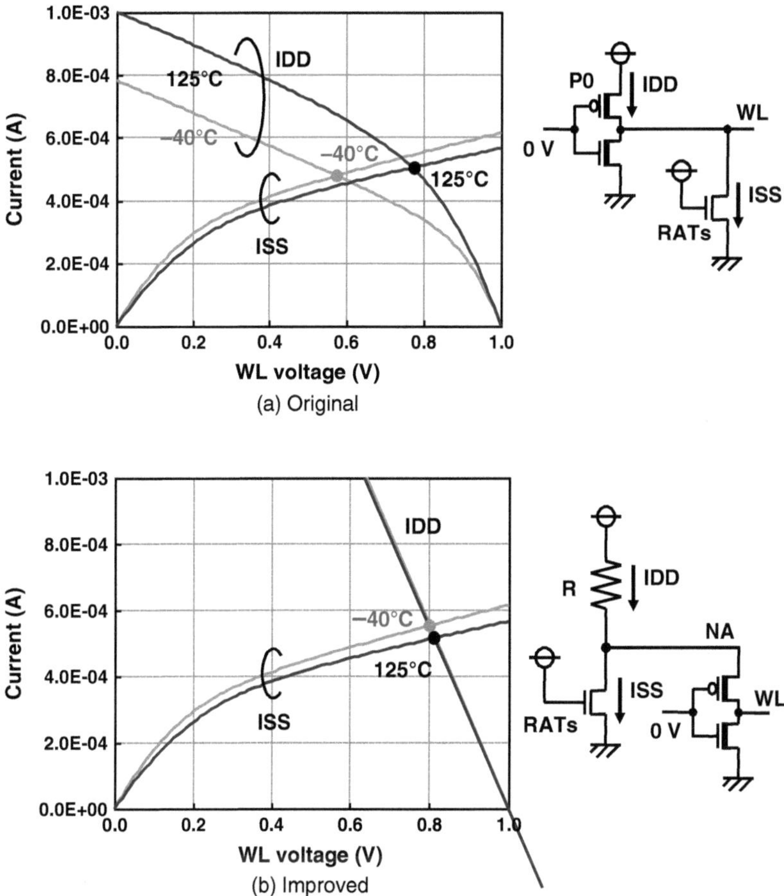

Fig. 5.26 Operating analysis of read assist circuit by SPICE simulation

In other words, positive CD shifts of polysilicon decrease the resistivity of R0 and R2, while negative CD shift of diffusion increases the R1 resistivity at the SS corner (see Fig. 5.29). Relatively lowered resistance of R2 rather than R1 contributes to pull down the NB node, which prevents NA node from lowering excessively. As a result, we can realize a higher voltage level of the V_{WL} at the SS condition compared to the case that we apply a simple pull-down transistor for RAT.

Now that we found the effectiveness of our RAC with the GC qualitatively, we verify its validity by simulating the V_{WL}. Figure 5.30 is the simulated result of the V_{WL} against NMOS V_{th} variation or process variation. To see the advantage of our RAC with the additional GC, we also show the result of the RAC without the GC. In the NMOS slow corner (SF, SS), the V_{WL} with GC is higher than that without GC, so that the degradation of cell current will be suppressed. Conversely in the NMOS

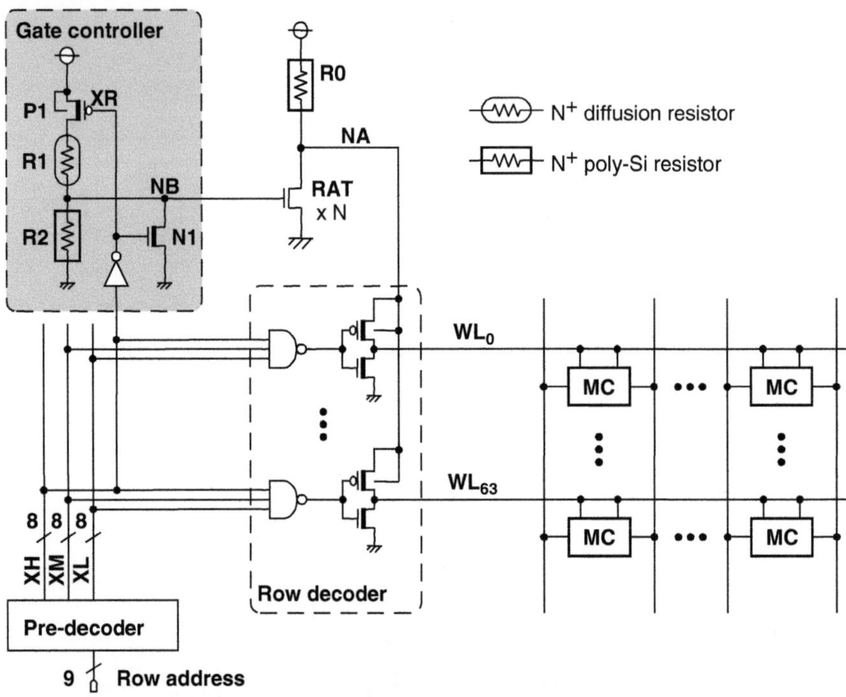

Fig. 5.27 Practical read assist circuit-enhanced sensitivity of process variation

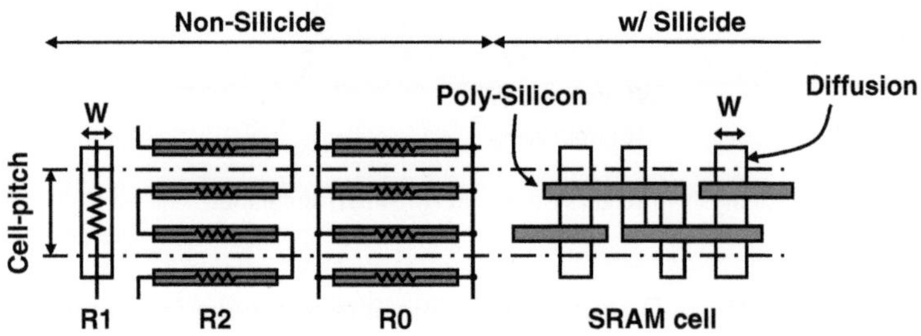

Fig. 5.28 Layout of each passive resistance for the proposed RAC

fast corner (FS, FF), the V_{WL} with GC is lower than that without GC, so that the read stability SNM will be improved. Figure 5.31 shows the simulated waveforms focusing on WL activation. In the inactivated blocks of rows, the node NA voltage level V_{NA} stays at VDD. When the node XR voltage level V_{XR} comes down to low level, the node NB voltage level V_{NB} rises to intermediate voltage level fixed by R1 and R2. Then the V_{NA} is dropped by the pull-down RATs. The substantial capacitance of the node NA makes its voltage drop gradually. Thus, it is found that the RAC does not affect the rising speed of the WL voltage.

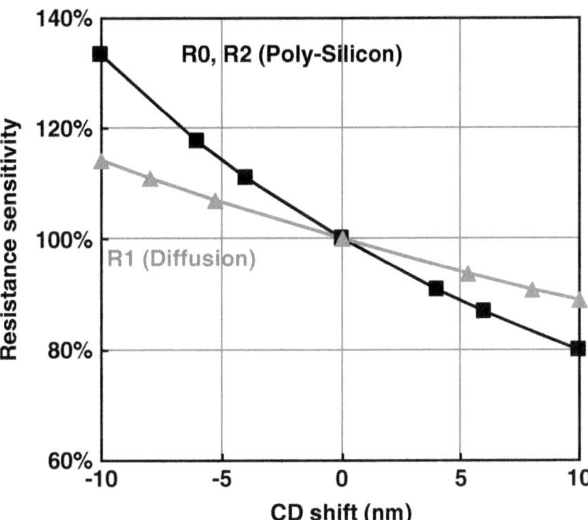

Fig. 5.29 Resistance sensitivities depend on critical dimension (CD) shift

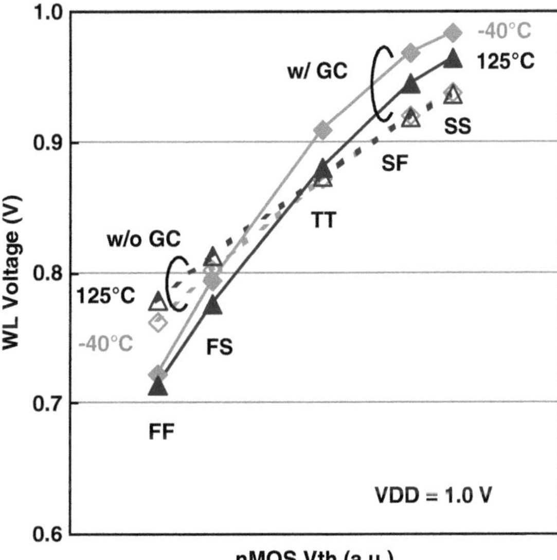

Fig. 5.30 Comparison of WL voltage (V_{WL}) in RAC with enhanced gate controller (GC) and without GC

5.3.3 Variation Tolerant Write Assist Circuits

Figure 5.32 shows the write assist circuit (WAC). Lowering the voltage level of power line in the memory cell array (ary-VDM) is one of the effective ways of ensuring the SRAM write margin [5.5, 5.6]. The capacitive WAC makes use of the capacitance ratio between the ary-VDM (C_{av}) and the additional dmy-VDM (C_{dv}) [5.5]. The dmy-VDM is wired by fourth metal layer upper to the cell array in each column individually. To enhance the write margin against the increasing variation

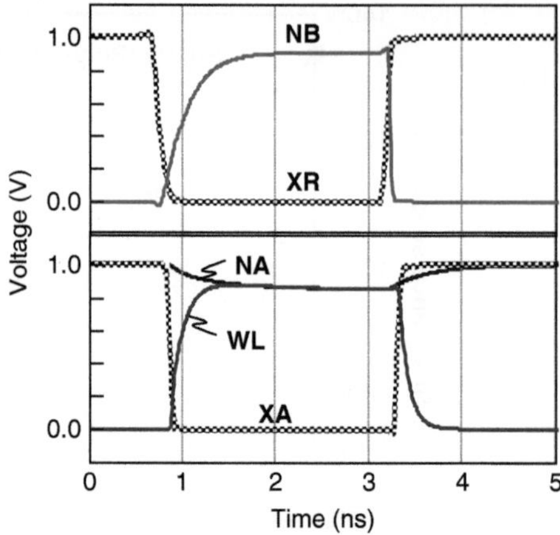

Fig. 5.31 Simulated waveform of proposed RAC

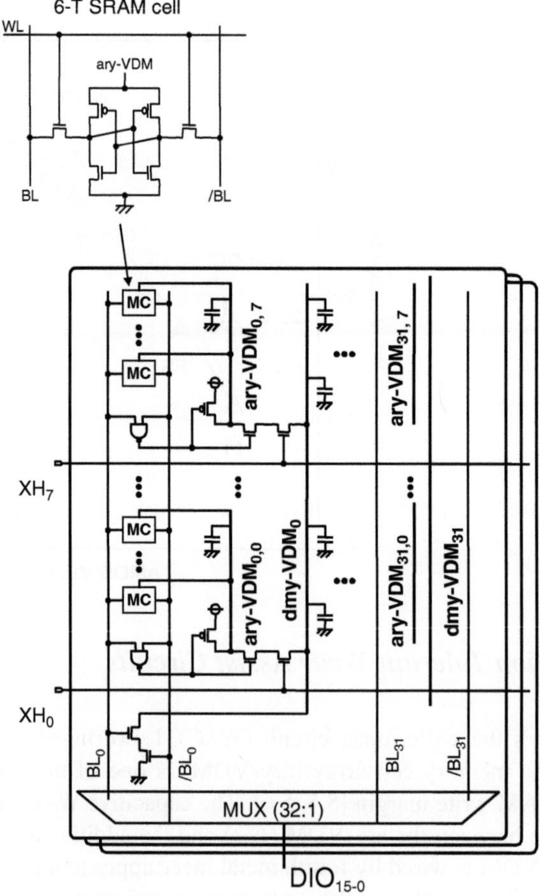

Fig. 5.32 Write assist circuit (WAC) improving write ability

accompanied by the scaling, it is necessary to lower the voltage of ary-VDM immediately, which requires a small C_{av}/C_{dv}. If the C_{dv} is enlarged, it degrades the speed and power of the writing operation. In Fig. 5.32, the signals of XH_0 and XH_7 indicate the most significant predecode signal of the row decoder to select one of the segment arrays. Each two-input NAND gate is connected to the bitline (BL) pair in each column of the segment arrays. In the NOP states where all BL pairs are highly pre-charged, all ary-VDM lines are connected to the power source line through the pull-up PMOS transistors, and all dmy-VDM lines are connected to the ground line through the stacked pull-down NMOS transistors. In the read state, although one of the BL pairs is slightly lowering in each column by flowing the cell current into the activated memory cell, the voltage levels of the BLs are almost high level, so that all ary-VDM and dmy-VDM keep the high level and low level, respectively. In the write state, as one of the BL pair is forced to low, both ary-VDM and dmy-VDM lines are put into a floating state. Consecutively, when the corresponding XH signal is activated, the corresponding ary-VDM line falls to the intermediate voltage level determined by C_{av}/C_{dv}. In Fig. 5.32, the number of column multiplex, data I/O, and divided segment are 32-column, 16-bit, and 8-segment, respectively, and the number of memory cells in each column in the segment is 64-cell as demonstrated in the following section.

Figure 5.33 shows the simulated waveform of the ary-VDM lowering in several cases of the segment division number (#div = 1,2,4,8). In this simulation, it is assumed that the C_{dv} is a constant value determined by the wire length of fourth metal of the 512 memory cells (#row = 512). The simulation condition is under the room temperature (RT) and 1.2 V typical supply voltage. In the case of #div being equal to 1, the voltage level of ary-VDM becomes 1.13 V from 1.2 V (-0.07 V), while it becomes lower 0.68 V (-0.52 V) in the case of #div being equal to 8 due

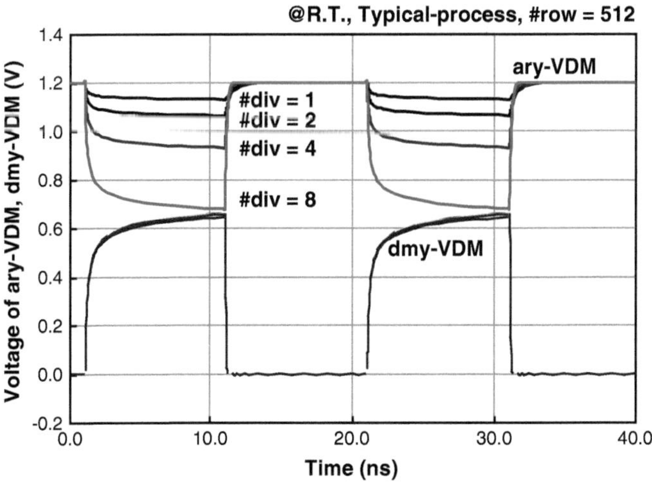

Fig. 5.33 Simulated waveform of the ary-VDM and dmy-VDM in the write status

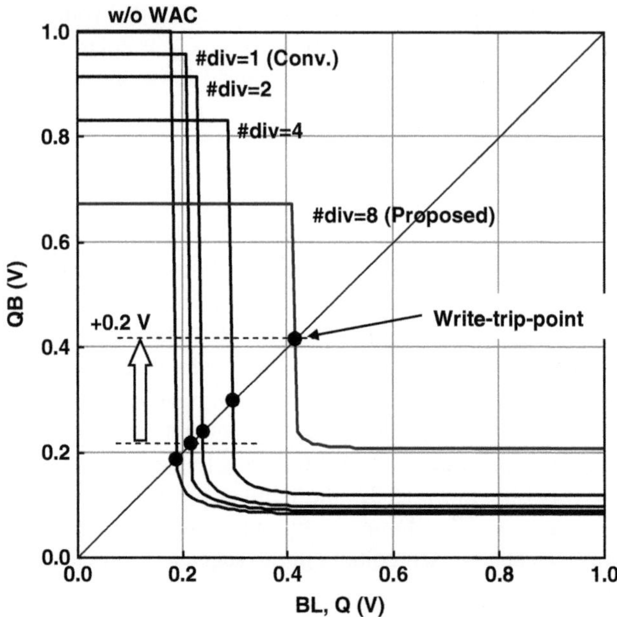

Fig. 5.34 Comparison of the write ability by DC simulation result of the write-trip-point

to the small $C_{\text{av}}/C_{\text{dv}}$. According to V_{th} curve simulation [5.4], to ensure the write margin in our 45-nm CMOS technology, the voltage level of ary-VDM should be 0.7 V at VDD = 1.0 V operation of the worst case, so that we divided the ary-VDM into eight as shown in Fig. 5.32.

Figure 5.34 shows the dc simulation result of improvement of write ability defined by write-trip-point [5.7]. The voltage levels in cases of #div = 1,2,4, and 8 of internal nodes Q and QB of a memory cell (as shown in Fig. 5.21) are plotted when one of the input of the BL pair (BL) is changed from 1.0 to 0.0 V and the other BLB is forced to 1.0 V of the pre-charge level, constantly. The initial condition is set Q equal to low level and QB equal to high level. As the voltage levels of BL and node Q become lower, the voltage level of node QB is flipped, and the cross point of the input BL (Q) and output QB is defined as the write-trip-point. The higher the voltage level of the write-trip-point, the higher the write ability; hence that means it is easy to flip the data. After the data flip, the voltage levels of QB in each case of #div becomes corresponding to lowered voltage level of the ary-VDM determined by $C_{\text{av}}/C_{\text{dv}}$ as mentioned before. It is found that the write-trip-point of proposed #div = 8 is 0.41 V at the worst condition, which is improved by 0.2 V compared to the conventional WAC (#div = 1).

The proposed divided WAC contributes to the suppression of the power overhead in the write operation compared to the conventional WAC [5.5]. Figure 5.35 shows the estimated power reduction in a write cycle at the memory cell array (MAT) depending on the number of the segment division. The referenced power is without

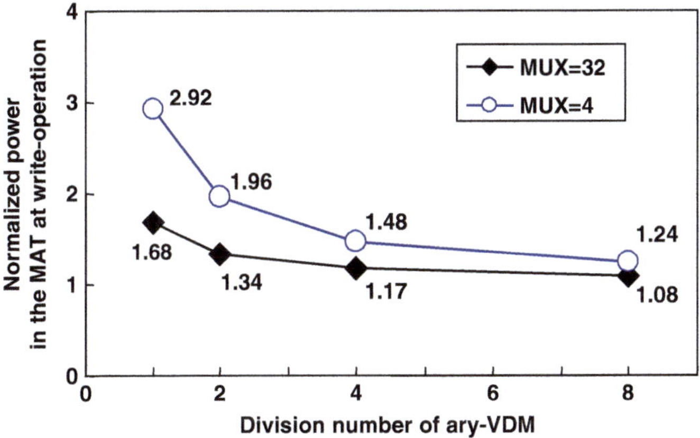

Fig. 5.35 Estimation of power reduction in the write cycle

WAC of column multiplex equal to 32, in which the power is dissipated by one full swing BL and the other 31 read cell currents by half-selected cells. The conventional WAC has 68% power overhead because the large capacitances of ary- and dmy-VDM need to be charged and discharged every write cycle. In contrast, the proposed eight-divided WAC has only 8% power overhead due to the small capacitances of ary- and dmy-VDM. The other case of column multiplex equal to 4, which is frequently used in embedded small SRAM macros of SoCs, is also plotted. It is found that the write power is further suppressed by dividing the ary-VDM.

5.3.4 Simulation Result

Figure 5.36 shows the practical simulated waveforms of a 512-kb SRAM macro in the read and write cycle in the best condition (FF-process, 1.4 V, 125°C) and worst condition (SS-process, 1.0 V, −40°C). The simulated clock cycle is 4.0 ns (250-MHz operation). Note that the first and the third clock cycles are in the write state, while the second and the fourth clock cycles in the read state. Owing to our proposed RAC during both in the read and in the write state, the V_{WL} is decreased reflecting the difference of the process condition: 30% reduction in the FF condition, while 8% reduction in the SS condition. On the other hand, during the write cycle, the V_{AV} at the SS (worst) condition is lowered from 1.0 to 0.75 V immediately after the BL starts swinging, which satisfies our requirement to obtain enough write margin against the variability. It should be commented that in the worst condition, both the voltage levels of dmy-VDM and ary-VDM are not equalized due to the NMOS pass-gate as shown in Fig. 5.32. We can avoid this voltage difference only by introducing a PMOS pass-gate. However, the equalization of these nodes makes the V_{AV} level lower, which results in the degradation of the retention margin of unselected memory cells in the activated column. The issue of the retention margin should be paid attention especially in the case of lower supply voltage. In addition,

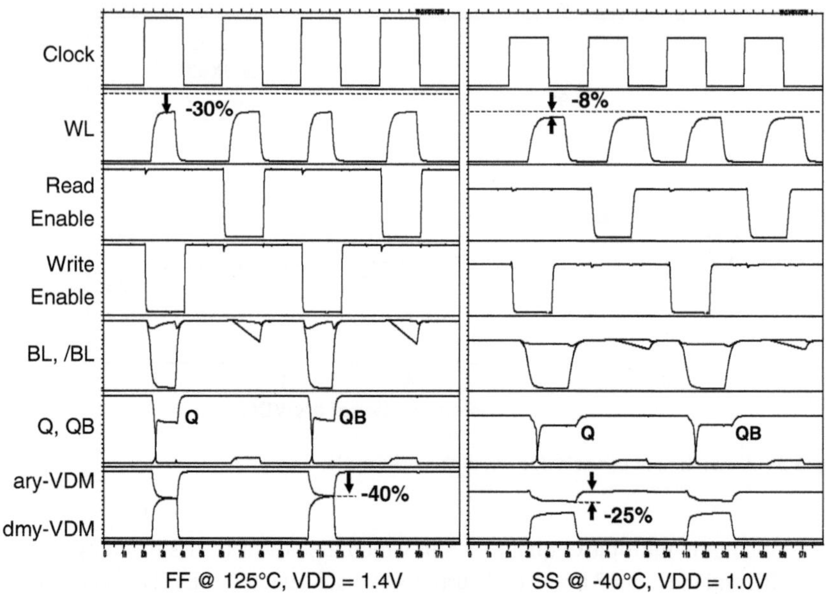

Fig. 5.36 Simulated waveform in read and write cycle

introduction of the extra PMOS leads to an area penalty. In this way, from the viewpoint of retention margin and the area penalty, we have adopted a single NMOS pass-gate between ary-VDM and dmy-VDM.

5.3.5 Fabrications and Evaluations in 45-nm Technology

The high-density and normal 6T SRAM cells are designed and fabricated using our 45-nm bulk CMOS technology with the SiON gate. The high-density cell ($0.245\,\mu\mathrm{m}^2$) has been shrunk by 50% while the normal one ($0.327\,\mu\mathrm{m}^2$) by 35%, providing that the SRAM cell size in 65 nm technology node is $0.49\,\mu\mathrm{m}^2$ [5.5]. Figure 5.37 shows the die photograph of the test chip for the normal-density SRAM cell. It consists of two conventional 256-kb SRAM macros and two proposed 256-kb ones. The physical layout size of the latter normal-density SRAM macro is $550 \times 305\,\mu\mathrm{m}^2$. The area penalty of the read and WAC are 7% and 9% for normal and high-density SRAM macros, respectively. The bit density of each SRAM macro is $1.57\,\mathrm{Mbits/mm}^2$ and $2.12\,\mathrm{Mbits/mm}^2$ for normal and high-density SRAM, respectively. The feature of fabricated SRAM macros is summarized in Table 5.1.

Figure 5.38 shows the relationship between minimum operation voltage and the process variations in the temperature between $-40°\mathrm{C}$ and $125°\mathrm{C}$ for the normal 512-kb SRAM macro. The dotted line represents the minimum operation voltages without the assist circuit, while the solid one that with the assist circuit. It is found

Fig. 5.37 Die photograph of a test chip

Table 5.1 Feature of the fabricated SRAM macro

Technology	45-nm (hp65) LSTP bulk CMOS	
Macro configurations	16 bits × 16-k words (MUX = 32)	
Memory cell arrays	512 rows × 256 columns × 2 MATs	
6T SRAM cell size	$0.327\,\mu m^2$	$0.245\,\mu m^2$
Bit density	$1.57\,Mbit/mm^2$	$2.12\,Mbit/mm^2$
Area overhead of assists	7%	9%
Access time at RT, 256 Kb	2.2 ns at 1.2 V	2.5 ns at 1.2 V
Standby leakage at RT	15 pA/cell at 0.7 V	26 pA/cell at 0.7 V

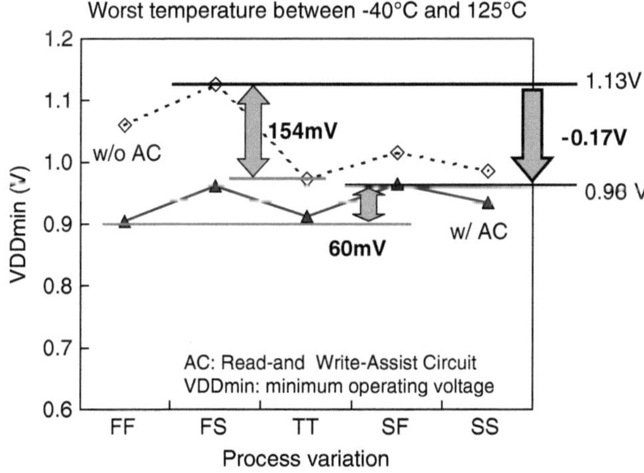

Fig. 5.38 Measured minimum operating voltage (VDD_{min}) of 512-kb SRAM at worst temperature

that the VDD_{min} in each process conditions (FF, FS, TT, SF, SS) is improved by read and WACs. The VDD_{min} is lowered from 1.13 to 0.96 V, achieving 170 mV improvement at worst process and temperature condition. Moreover, it should be also noted that the distribution of VDD_{min} in accordance with the process variations is suppressed from 154 to 60 mV by applying the assist circuit. These results indicate that the SRAM with the assist circuit has improved immunity against process and temperature variations.

5.4 Dual-Port Array Design Techniques

5.4.1 Access Conflict Issue of Dual-Port SRAM

Figure 5.39 shows memory cell circuits for single-port and dual-port SRAM. The standard single-port SRAM cell presented in Fig. 5.39a comprises six transistors: two pull-up PMOSs (load-PMOS), two pull-down NMOSs (drive-NMOS), and two transfer NMOSs (access-NMOS). The single-port SRAM realizes either read operation or write operation, so that its operation is often denoted as "1RW."

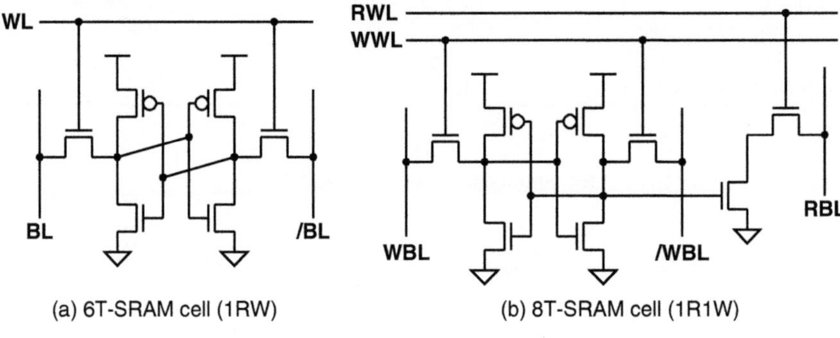

(a) 6T-SRAM cell (1RW)   (b) 8T-SRAM cell (1R1W)

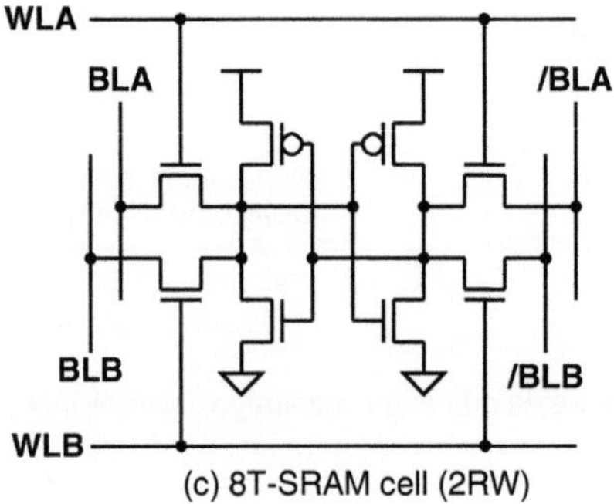

(c) 8T-SRAM cell (2RW)

Fig. 5.39 SRAM memory cell circuits

Normally, as presented in Fig. 5.39b, c, two major types of memory cells are used for the dual-port SRAM. Although both memory cells have eight transistors in common, their function differs greatly. Figure 5.39b portrays the one-read/one-write (1R1W) type DP-SRAM cell, in which only one of the two ports is allowed for read operation. This 1R1W memory cell has stable read operation, but its single-ended read-bitline (RBL) structure might have an impact of access time degradation, unfortunately, because of the RBL full swing. Figure 5.39c portrays the two read–write (2RW) type of 8T SRAM memory cells. In this type of dual-port memory cell, both ports are available for reading and writing, which indicates that the 2RW type of memory cell can also operate as a 1R1W, although the 1R1W type of memory cell cannot operate as a 2RW. In this way, the 2RW type of 8T DP-cell has more access flexibility. Hereafter, this type of DP-SRAM is specifically addressed in this study.

Figure 5.40 depicts the variety of the access situations of the 2RW dual-port SRAM when both ports are enabled simultaneously. The memory cell array with activated 8T-cells, word lines (WLA, WLB), and bitlines (BLA, BLB) is shown simply. The buffers of both sides of memory cell array designate the addressed WL drivers of both ports. Figure 5.40a depicts a situation in which a different row and

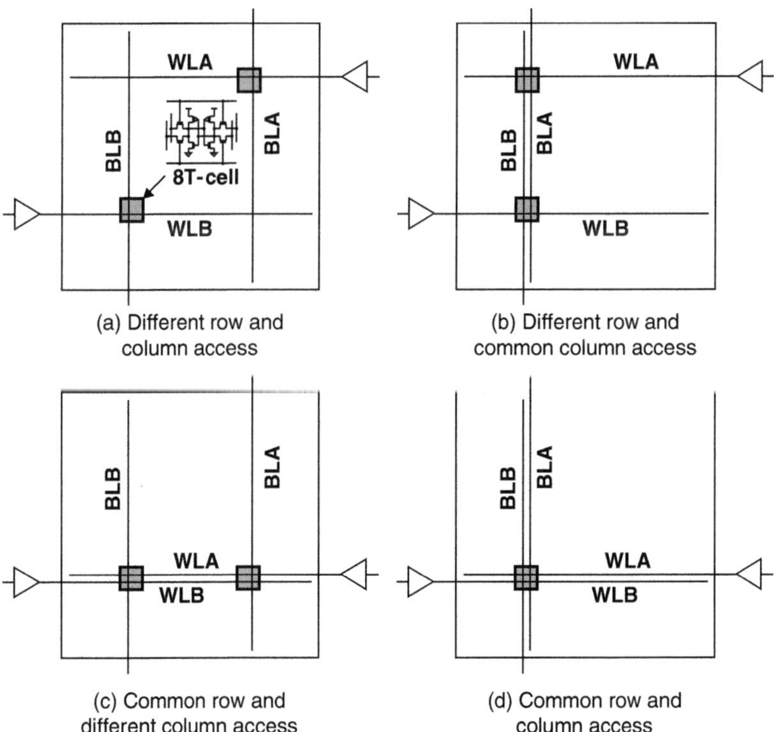

Fig. 5.40 Assortment of the access modes of the dual-port SRAM

column are accessed from both ports, which are designated independently by each address input. Figure 5.40b shows the different row and common column access situation. These two situations have no issues in terms of the access conflict of both ports because each selected memory cell, of which either WLA or WLB is activated, operates as a single-port access. Figure 5.40c, d, respectively, shows the common row and different column access, and the common row and common column access. In these common row access situations, the cell stability must be considered as the worst case for reading because the two enabled word lines affect the static noise margin (SNM) degradation for all memory cells along with the selected row. Both ports operate as reading; also, one port operates as writing or both ports operate as writing. Therefore, the write stability is also considered as the worst case of the selected memory cell. The read stability is still considered in writing operations because the half-selected (selected row and unselected column) memory cells are equal to reading situations even if one or both ports are performing a writing operation. In general, if the writing data are different (namely the opposite data) from both ports, absolutely consistent address access for a writing operation from each port, as presented in Fig. 5.40d, is inhibited because of the abnormal leakage current flows in the accessed memory cell. Still, the simultaneous reading operation or reading and writing operations from both ports are frequently required from the system. Therefore, the conventional DP-SRAM design must satisfy such a worst case access situation: the size of 8T DP-cell necessarily becomes large because of increasing gate width of drive-NMOSs to improve the cell stability.

Figure 5.41 shows simulated butterfly curves of the SNM for the 8T DP-cell. As described earlier, the 8T DP-cell has two different SNM values depending on the access situation: one is a common access situation in which two word lines (WLs) within the same row are selected; the other is a different access situation in which two WLs in two different rows are selected. In the common access situation shown in Fig. 5.40c, d, both WLs are activated. Therefore, the electrical β ratio of the 8T DP-cell is expressed as $\beta_{ND1}/(\beta_{NA1} + \beta_{NA2})$. Here, β_{ND1}, β_{NA1}, and β_{NA2}, respectively, indicate the coefficients of source-drain currents of the drive-NMOS, the access-NMOS for port A, and the access-NMOS for port B. On the other hand, as for the different access situation, the corresponding β ratio becomes β_{ND1}/β_{NA1} or β_{ND1}/β_{NA2} because of single activation of the WL. In general, a lower β ratio reduces the read stability, SNM, which indicates that the SNM in common access situation must be discussed for the worst case design of the 8T DP-cell.

5.4.2 Circumventing Access Scheme of Simultaneous Common Row Activation

Figure 5.42 presents the fundamental concept of the proposed DP-SRAM access scheme. For convenience, it is defined that port A, which is connected to the pair of BLA and /BLA, is primary, whereas port B, which is connected to that of BLB and /BLB, is secondary. In the secondary port B, the row address comparator (RAC)

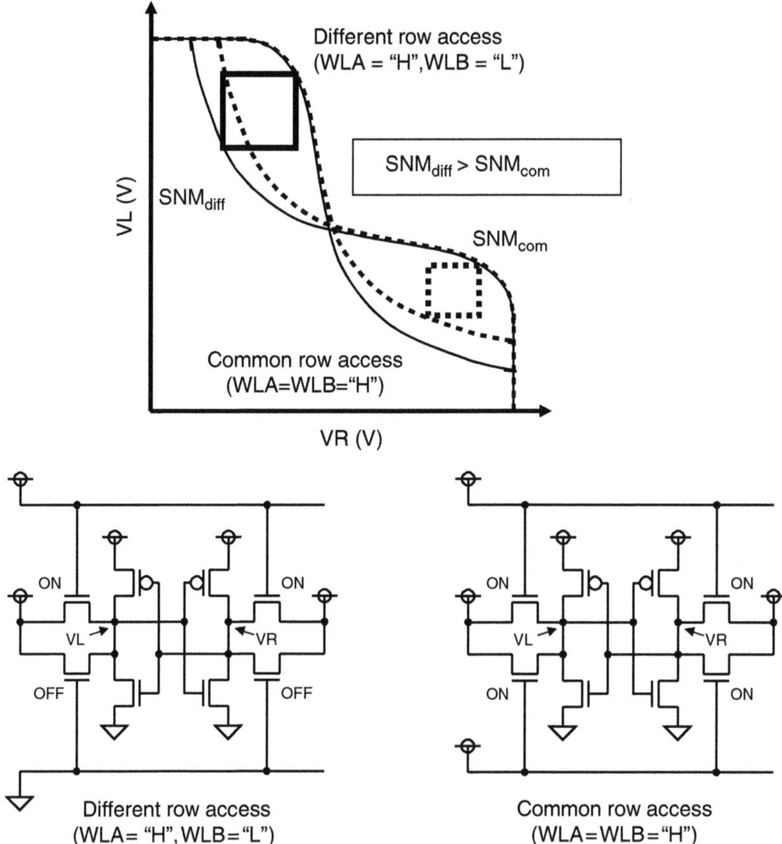

Fig. 5.41 Butterfly curves and static noise margin of the DP-8T-cell for both common row access and different row access

and the bitline shifting circuitry (BSC) are introduced. Figure 5.43 expresses more detailed operations depending on the access mode.

The implemented circuitry in a test chip design is portrayed in Figs. 5.44 and 5.45. Figure 5.43a shows that the address input signal AA <> activates WLA in the m-th row (WLAm), whereas the AB <> activates WLB in the n-th row (WLBn), which means they have different access modes. In this condition, the RAC is designed to the output "H" level so that the DP-SRAM as a whole is expected to realize a standard read or write operation. Once the AA <> and the AB <> select the WLs in a common row, as presented in Fig. 5.43b, the row decoder for port B is disenabled because of the RAC. Consequently, only the WLAn is accessible to the memory cell. Simultaneously, the "L" level generated by the RAC (see also Fig. 5.44) modifies the connection of secondary port B from the pair of BLB to that of BLA, making it possible to read data stably without SNM degradation. In other words, this scheme circumvents the common access mode.

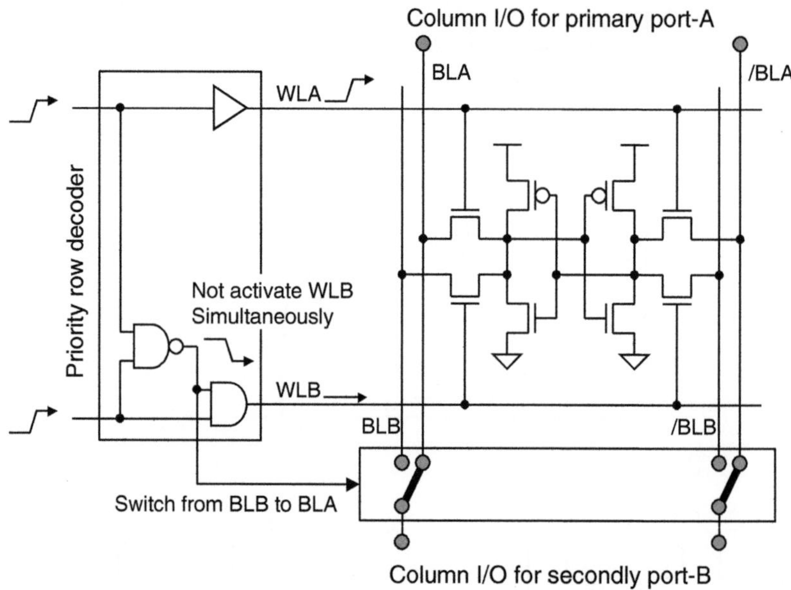

Fig. 5.42 Concept of proposed circumventing simultaneous common row access

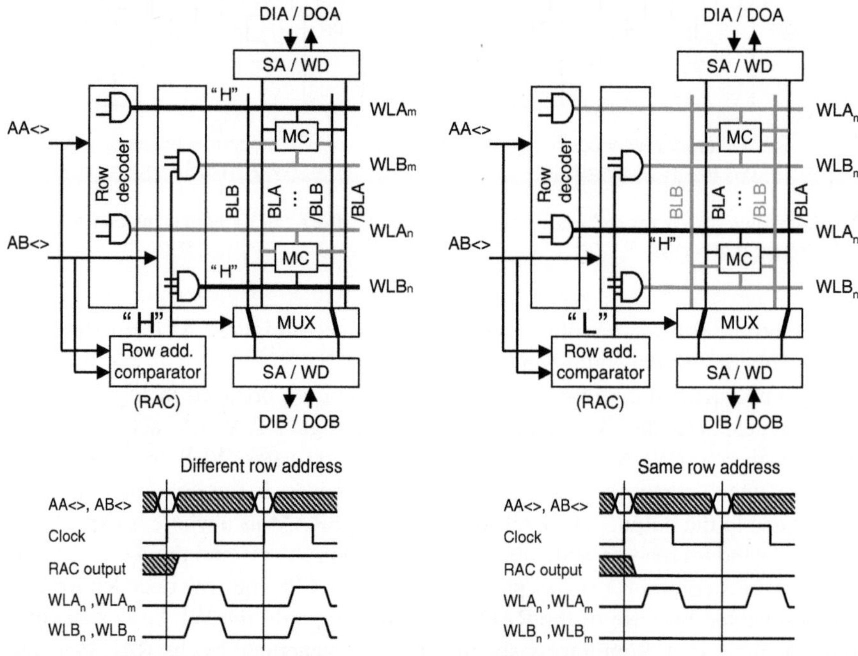

Fig. 5.43 Block diagram and timing chart of the proposed access scheme

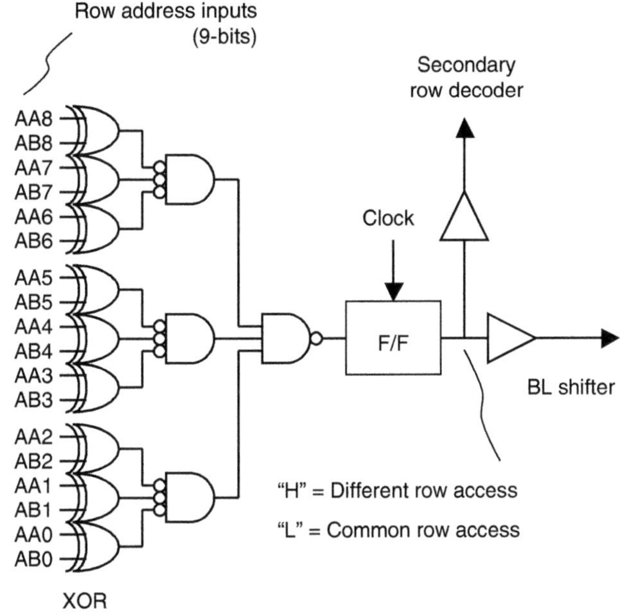

Fig. 5.44 Circuit of the row-address comparator (RAC)

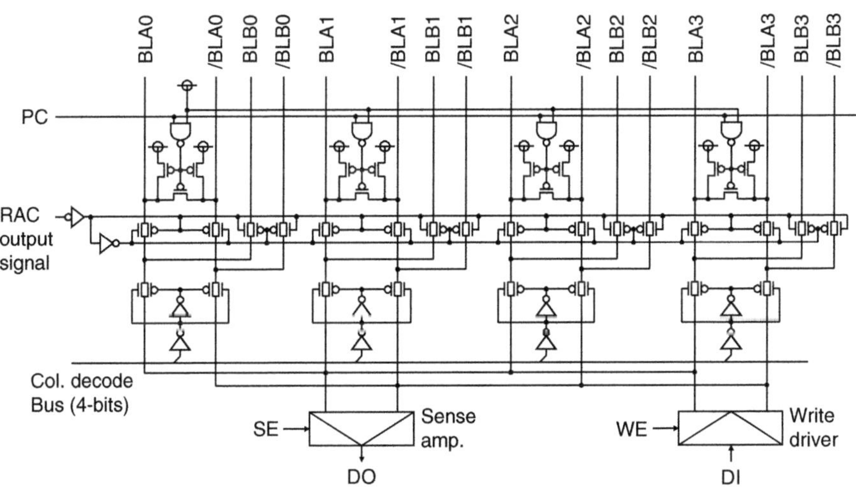

Fig. 5.45 Circuit of the bitline shifter for secondary port

For that reason, it is possible to reduce the drive-NMOS transistor width, which contributes directly to the reduction of the DP-SRAM unit cell area. In addition, this circuitry has a strong effect on the write operation. In fact, the common access mode becomes a critical problem in the write operation because the read operation takes place in unselected columns, which means that the data to be stored might be flipped during writing. However, the proposed scheme keeps the WLBs at "L" levels as well as in the write operation. Consequently, whenever the common access mode occurs, this type of error can be avoided safely. In this way, the fatal risk associated with the specific operation in the DP-SRAM can be circumvented. Furthermore, it is noteworthy that the introduction of the additional circuitry is compensated by the reduction of the cell area of a unit DP-SRAM.

5.4.3 8T Dual-Port Cell Design

Figure 5.46 shows scaling trends of embedded SRAM cell size of 6T SRAM (for a 1RW single port) and 8T SRAM (for a 2RW dual-port). The cell size of 6T SRAM shrinks by half as one technology node advances. Conventionally, the 8T DP-cell sizes are more than twice as large as 6T SP-cell sizes up to 130-nm technology. A new elongated 8T DP-cell layout was proposed; its cell size was $2.04\,\mu m^2$, which is only 1.63 times larger than 6T SP-cell of the $1.25\,\mu m^2$ in 90 nm technology [5.8, 5.9]. According to the scaling trend, both the 6T SP-cell and 8T DP-cell sizes become approximately half, which are 0.61 and $0.99\,\mu m^2$, respectively, in 65 nm technology with the same layout topology [5.10]. The new access scheme is applied to achieve a smaller cell beyond the scaling trend. In addition, aggressive shrinkage is obtainable by making active diffusion and polysilicon gates into regular polygons from the design for manufacturability (DFM) perspective, providing improved printability of the cell layout. As a result, the proposed thin 8T DP-cell size is $0.71\,\mu m^2$, which is 30% smaller than a normal 8T DP-cell and is only 1.44× the cell size of an advanced high-density 6T SP-cell [5.11].

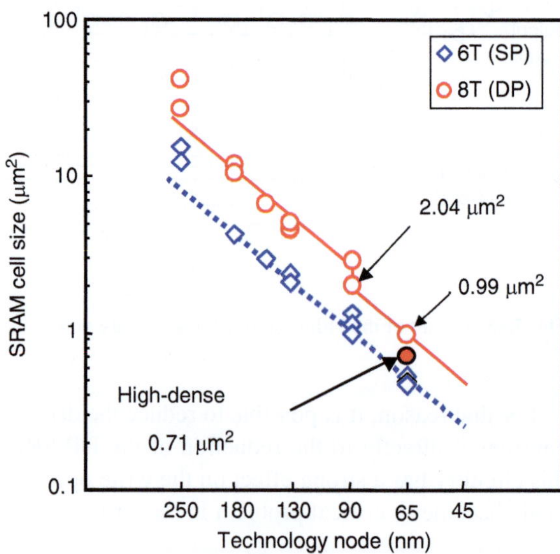

Fig. 5.46 Scaling trend of the SRAM memory cell size

Beyond 100-nm technology, the major memory cell layout of 6T SRAM becomes a wide and thin rectangle type, which includes two well-bounded regions. Extending the same layout topology, the conventional 2RW type of an 8T SRAM cell layout [5.8] was also a thin rectangle type similar to a 6T SRAM cell. The proposed high-density 8T DP-cell layout is based on these wide and thin rectangle types. Figures 5.47 and 5.48 show the layout and an SEM image of the proposed 8T DP-cell using 65 nm LSTP CMOS technology. As with a conventional 8T DP-cell, four shared contacts of tungsten plugs connect the polysilicon gate and diffusion region directly to achieve a smaller cell size. In terms of front-end-of-line (FEOL), the cell width (x direction) can be shrunk aggressively because the transistor width of drive-NMOS can be reduced to about half that of the normal cell. Regarding the back-end-of-line (BEOL), however, no scaling down occurs in the x direction because the second metal tracks consisting of BL pairs, WL islands, and power line are almost completely occupied, even for a conventional 8T DP-cell.

To resolve this BEOL bottleneck, the layers of BLs, WLs, and the power line to upper layers are changed in each, as presented in Fig. 5.48b. The BLs and power line run with the third metal layer in vertical direction and the WLs run with the fourth metal layer in the horizontal direction. The ground-line maintains a second metal layer, but it is connected directly with both sides in each cell in a zigzag wire, as in a snake pattern. Consequently, the required second metal tracks are reduced to seven from nine; the cell width is then determined using FEOL, not BEOL.

In this design, the electrical β ratio is reduced to one, which minimizes the 8T DP-cell width in x directions, i.e., the $\beta_{ND1} = \beta_{NA1} = \beta_{NA2}$, which is the same ratio as that of the 6T SP-cell [5.10]. For that reason, the regions of n-type active diffusions and polysilicon gates become a straight polygon pattern, which presents advantages from the DFM perspective. It is lithographically friendly or robust against misalignment of mask steps because of the reduction of the corner round shapes. Therefore, the minimum dimensions of FEOL can be reduced aggressively without yield loss. Regarding concerns about the read stability attributable to the small electrical β ratio, the topic is discussed in the following two subsections.

5.4.4 Simulated Butterfly Curves for SNM

Next, the read stability of the proposed 8T DP-cell is discussed. Figure 5.49 shows the simulated butterfly curves both of conventional DP-SRAM cell and the proposed UHD-8T-SRAM cell in 65-nm technology. The data show the process under typical conditions: 1.2 V supply voltage and room temperature. Conventional 8T SRAM must be considered the worst case of the common row access situation. On the other hand, the proposed 8T DP-SRAM is considered to be the case in which either WLA or WLB is activated like a 6T SP-SRAM. The DC simulation result shows that the SNM values are 186 and 194 mV, respectively, for conventional and proposed 8T DP-SRAMs. In spite of the small electrical β ratio, the SNM of the proposed

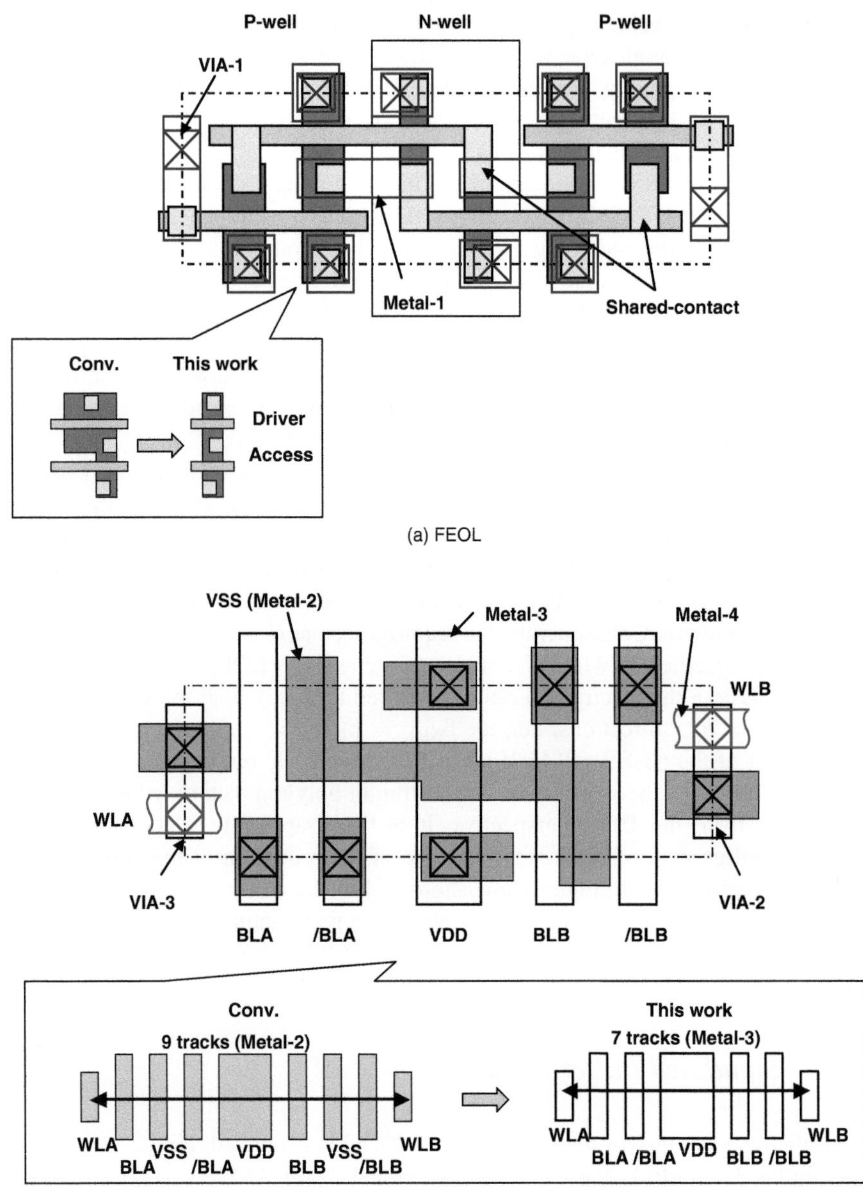

Fig. 5.47 The 8T-DP-cell layout

UHD-8T-SRAM cell is slightly larger than the conventional DP-SRAM cell under typical conditions because that the V_{th} of small access transistor of conventional unit cell is lower as a result of the reverse narrow effect. Meanwhile, the V_{th} of access transistor of the proposed cell is almost the same as that of the driver transistor [5.10].

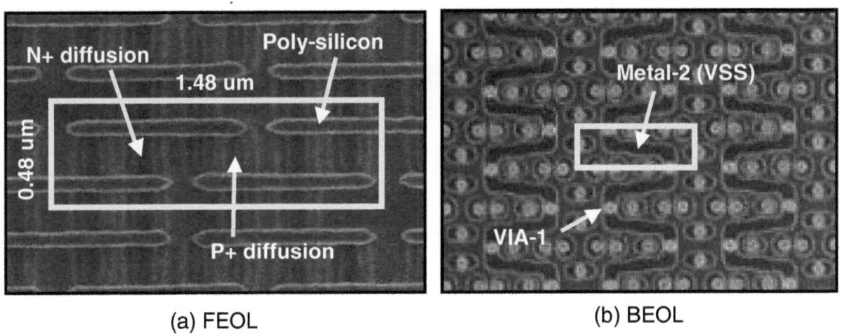

(a) FEOL   (b) BEOL

Fig. 5.48 SEM image of 8T-DP-cell after poly etching and metal-2 damascening

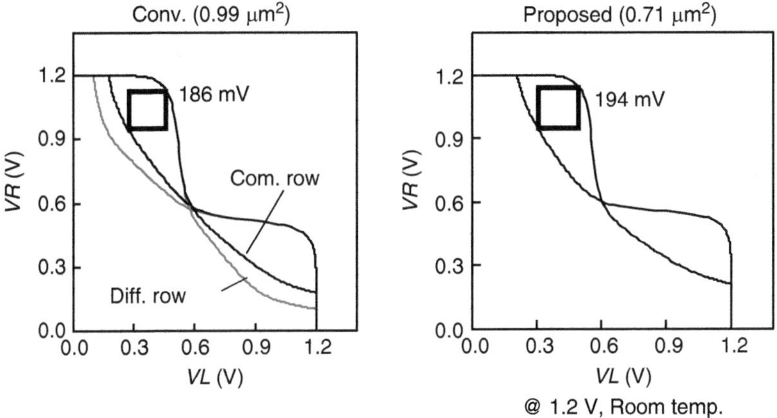

Fig. 5.49 Measured SNM for conventional and proposed 8T-DP-cells

5.4.5 Cell Stability Analysis

The stability of proposed DP-SRAM cell is verified by considering the global and the local V_{th} variation. The global variation means the inter-die variation, which results from variation of the gate length, gate width, gate oxide thickness, and dopant implantation. The local variation is the intra-die variation, which results from dopant fluctuation of channel and gate line-edge-roughness (LER). Figure 5.50 shows the result of stability analysis by V_{th} curve simulation [5.12] considering both global and local V_{th} variations. The read and write boundaries are solved using "the worst case model analysis." In this analysis, it is assumed that the total memory capability of DP-SRAM in one die is up to 1 Mbit. The temperature is −40 to 125°C; the supply voltage is 1.2 V ± 10%. As presented in Fig. 5.50, the read and write margin is sufficiently good for the global corner models FF, FS, SS, and SF, as well as a typical model CC. Here, FS means fast NMOS and slow PMOS; SF means slow

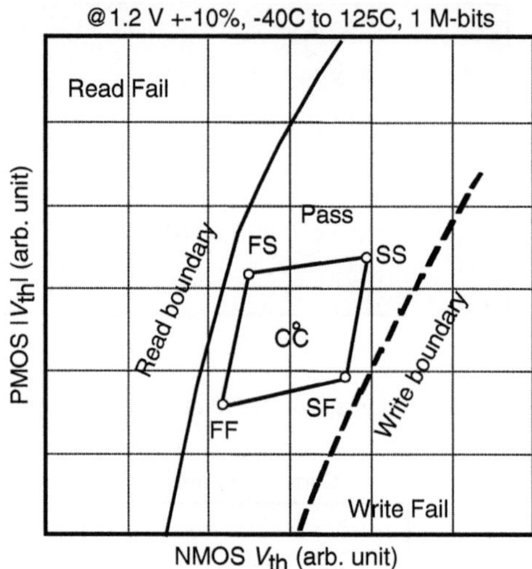

Fig. 5.50 Read-stability and write-ability analysis by V_{th} curve simulations

NMOS and fast PMOS. This simulation result shows that no yield loss pertains to mass production on account of DP-SRAM instability.

5.4.6 Standby Leakage

The small drive-NMOS transistor contributes not only to area but also to standby leakage reduction. Figure 5.51 presents a comparison of the simulated standby leakages of the reference $0.49\,\mu\text{m}^2$ 6T SRAM (SP) cell, the proposed $0.71\,\mu\text{m}^2$ 8T SRAM (DP), and the conventional $0.99\,\mu\text{m}^2$ 8T SRAM (DP) cell using 65-nm CMOS technology. For each cell, the total leakage current flow, which is the sum of the subthreshold leakage current, the gate-induced drain current (GIDL), and the gate leakage of all transistors are estimated. The typical standby leakage of the proposed DP 8T cell is 9.0 pA/cell at the 1.2 V supply voltage; the room temperature is reduced by 30% from that of the conventional 8T DP-cell. The increasing standby leakage of proposed 8T DP-cell is only suppressed to 1.4 times that of the 6T SP-cell because the leakage component of access-NMOS transistors was only twice.

5.4.7 Design and Fabrication of Test Chip

Test chips with eight embedded 32-kB DP-SRAM macros are designed and fabricated using 65-nm CMOS technology. Figure 5.52 shows a microphotograph of the $36.2 - \text{mm}^2$ test chip. The four macros at the right side are the proposed ultra-high-density (UHD) DP-SRAM, whereas the other four macros at the left side

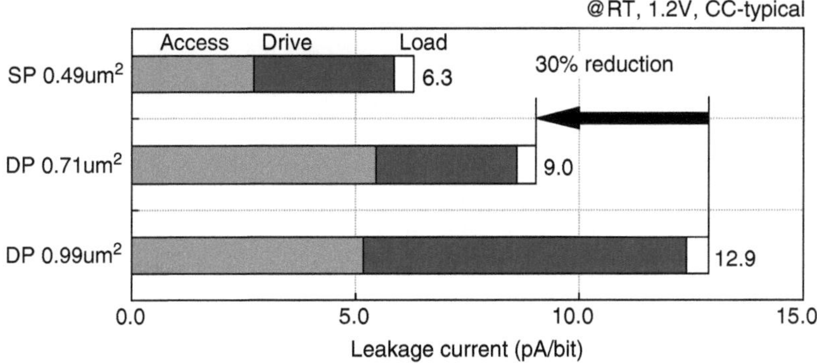

Fig. 5.51 Estimation of the standby leakage of the 8T-DP-cell by SPICE simulation

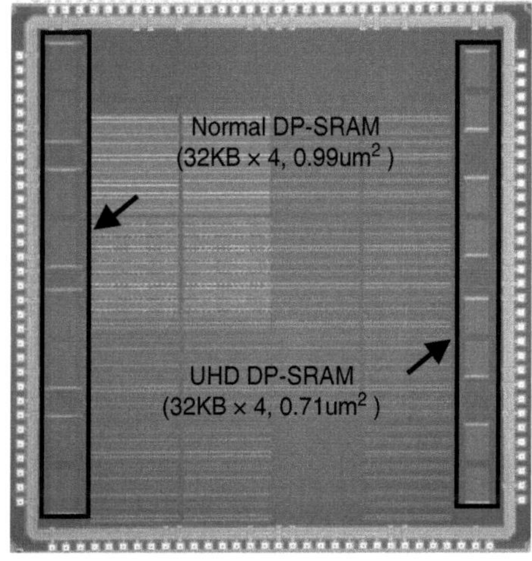

Fig. 5.52 Die photograph of a test chip

are normal DP-SRAM macros. Although the test chips were fabricated with eight metal layers, both conventional and proposed SRAM macros were implemented within four metal layers. Figure 5.53 presents the layout plot of the proposed 32-kB UHD DP-SRAM macros. Two row decoders for both port A and port B are placed exactly at the center of the macro, so that the memory cell array is divided into two cell arrays by the row decoder, thereby shortening the word line. The primary data I/O for port A are located at the upper side and secondary data I/O for port B are located at the opposite lower side. The BL shifter is inserted between the cell array and secondary data I/O is not inserted in the primary data I/O. The RAC is placed into the secondary address buffer region. The total cell array region is decreased by 30% because of the small memory cell compared to conventional one. On the other hand, the BL shifter and the RAC in the peripheral part are slightly larger by

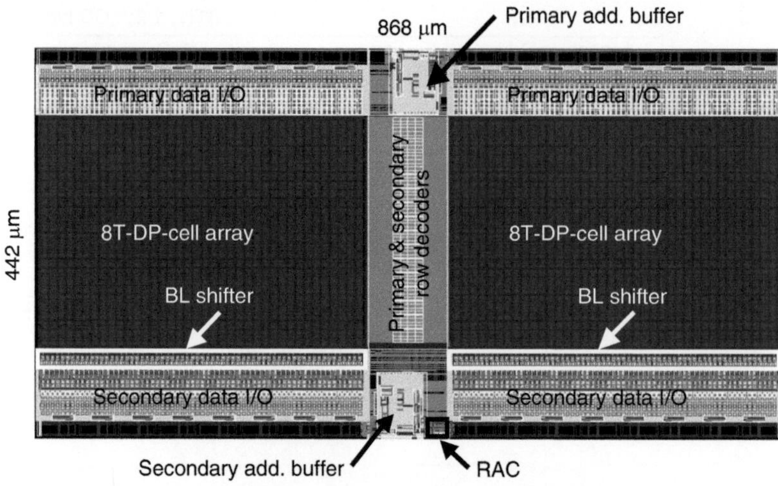

Fig. 5.53 Layout plot of fabricated 32-kB UHD-DP-SRAM macros

Table 5.2 Features of the fabricated SRAM macro

	Normal DP-SRAM	UHD DP-SRAM
Technology	65 nm (hp90) LSTP CMOS	
Configuration	16 bit × 16k word × 4 macro	
MAT size	512 row × 256 column × 2 MAT/macro	
Mux I/O	32	
Memory cell size	$0.99\,\mu m^2$	$0.71\,\mu m^2$
Physical macro size	$1,084\,\mu m \times 442\,\mu m$	$868\,\mu m \times 442\,\mu m$
Bit density	$534\,kbit/mm^2$	$667\,kbit/mm^2$
Read access time at 1.2 V	3.1 ns	3.0 ns
Standby leakage at 1.2 V	$27\,\mu A/Mbit$	$20\,\mu A/Mbit$

5%. The physical layout of the 32-kB macro is $868 \times 442\,\mu m^2$; the bit density is $667 - kbit/mm^2$, achieving 25% increment of bit density.

5.4.8 Measurement Result

All 32-kB macros are tested and confirmed as fully operational. Table 5.2 summarizes the test chip features. Figure 5.54 presents a typical shmoo plot depending on the supply voltage versus clock access time under room temperature conditions. It shows the measured SRAM macro functions of 0.8–1.44 V. The measured clock access time was 3.0 ns at typical supply voltage 1.2 V; the conventional access time is 3.1 ns, which suggests a lack of an access time penalty (see Fig. 5.53). The measured typical standby leakage of four 32-kB macros (totally 128 kB) including

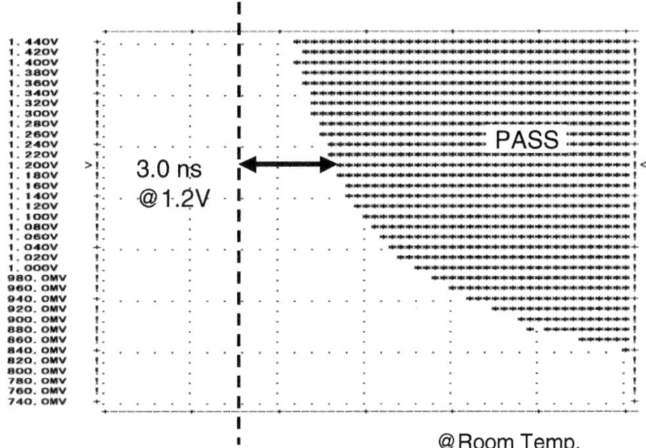

Fig. 5.54 Shmoo plot

both the cell array and peripheral was 20 μA, which was reduced by 27% from that of the conventional one because of the small drive transistor of the DP-SRAM cell.

References

5.1. S. Bharadwaj et al., A replica technique for wordline and sense control in low-power SRAM's. IEEE J. Solid-State Circuits **33**, 1208–1219 (1998)

5.2. E. Seevinck, F.J. List, J. Lohstroh, Static-noise margin analysis of MOS SRAM cells. IEEE J. Solid-State Circuits **SC-22**(5), 748–754 (1987)

5.3. K. Itoh, A.R. Fridi, A. Bellaouar, M.I. Elmasry, A deep sub-V, single power-supply SRAM cell with multi-Vt, boosted storage node and dynamic load, in*Symposium on VLSI Circuits, Digest of Technical Papers*, June 1996, pp. 132–133

5.4. Y. Tsukamoto, K. Nii, S. Imaoka, Y. Oda, S. Ohbayashi, T. Yoshizawa, H. Makino, K. Ishibashi, H. Shinohara, Worst-case analysis to obtain stable read/write DC margin of high density 6T-SRAM-array with localVth variability, in *ICCAD Digest of Technical Papers* (2005), pp. 398–405

5.5. S. Ohbayashi, M. Yabuuchi, K. Nii, Y. Tsukamoto, S. Imaoka, Y. Oda, T. Yoshihara, M. Igarashi, M. Takeuchi, H. Kawashima, Y. Yamaguchi, K. Tsukamoto, M. Inuishi, H. Makino, K. Ishibashi, H. Shinohara, A 65-nm SoC embedded 6T-SRAM designed for manufacturability with read and write operation stabilizing circuits. IEEE J. Solid-State Circuits **42**(4), 820–829 (2007)

5.6. K. Zhang, U. Bhattacharya, C. Zhanping, F. Hamzaoglu, D. Murray, N. Vallepalli, Y. Wang, B. Zheng, M. Bohr, A 3-GHz 70-mb SRAM in 65-nm CMOS technology with integrated column-based dynamic power supply. IEEE J. Solid-State Circuits **41**(1), 146–151 (2006)

5.7. R. Heald, P. Wang, Variability in sub-100 nm SRAM designs, in *ICCAD Digest of Technical Papers* (2004), pp. 347–352

5.8. K. Nii, Y. Tsukamoto, S. Imaoka, H. Makino, A 90 nm dual-port SRAM with 2.04 um² 8T-thin cell using dynamically-controlled column bias scheme, in *IEEE ISSCC Digest of Technical Papers*, February 2004, pp. 508–543

5.9. K. Nii, Y. Tsukamoto, T. Yoshizawa, S. Imaoka, Y. Yamagami, T. Suzuki, A. Shibayama, H. Makino, S. Iwade, A 90-nm low-power 32-kB embedded SRAM with gate leakage suppression circuit for mobile applications. IEEE J. Solid-State Circuits **39**(4), 684–693 (2004)

5.10. S. Ohbayashi, M. Yabuuchi, K. Nii, Y. Tsukamoto, S. Imaoka, Y. Oda, T. Yoshihara, M. Igarashi, M. Takeuchi, H. Kawashima, Y. Yamaguchi, K. Tsukamoto, M. Inuishi, H. Makino, K. Ishibashi, H. Shinohara, A 65-nm SoC embedded 6T-SRAM designed for manufacturability with read and write operation stabilizing circuits. IEEE J. Solid-State Circuits **42**(4), 820–829 (2007)

5.11. K. Nii, Y. Masuda, M. Yabuuchi, Y. Tsukamoto, S. Ohbayashi, S. Imaoka, M. Igarashi, K. Tomita, N. Tsuboi, H. Makino, K. Ishibashi, H. Shinohara, A 65 nm ultra-high-density dual-port SRAM with 0.71 μm^2 8T-cell for SoC, in *Symposium on VLSI Circuits Digest of Technical Papers*, June 2006, pp. 162–163

5.12. Y. Tsukamoto, K. Nii, S. Imaoka, Y. Oda, S. Ohbayashi, T. Yoshizawa, H. Makino, K. Ishibashi, H. Shinohara, "Worst-case analysis to obtain stable read/write DC margin of high density 6T-SRAM-array with local Vth variability," in ICCAD Digest of Technical Papers, (2005), pp. 398–405.

Chapter 6
Reliable Memory Cell Design for Environmental Radiation-Induced Failures in SRAM

Eishi Ibe and Kenichi Osada

Acronyms

CHB	CHecker Board
CHBc	CHecker Board complement
CORIMS	COsmic Radiation IMpact Simulator
DRAM	Dynamic random access memory
FPGA	Field programmable gate array
MBU	Multi-bit upset
MCBI	Multi-coupled bipolar interaction
MCU	Multi-cell upset
MNT	Multi-node transient
PCSE	Power cycle soft-error
SBU	Single bit upset
SEB	Single event burnout
SEFI	Single event functional interruption
SEGR	Single event gate rupture
SEL	Single event latchup
SER	Soft-error rate
SESB	Single event snapback

E. Ibe (✉)
Production Engineering Research Laboratory, Hitachi, Ltd., 292 Yoshida, Totsuka, Yokohama, Kanagawa 244-0817, Japan
e-mail: hidefumi.ibe.hf@hitachi.com

K. Osada
Measurement Systems Research Department, Central Research Laboratory, Hitachi Ltd., 1-280, Higashi-koigakubo, Kokubunji-shi, Tokyo 180-8601, Japan
e-mail: kenichi.osada.aj@hitachi.com

K. Ishibashi and K. Osada (eds.), *Low Power and Reliable SRAM Memory Cell and Array Design*, Springer Series in Advanced Microelectronics 31, DOI 10.1007/978-3-642-19568-6_6, © Springer-Verlag Berlin Heidelberg 2011

SET   Single event transient
SEU   Single event upset
SRAM   Static random access memory

Abstract In this chapter, current status of environmental radiation-induced failures in SRAM is introduced. Alpha ray-induced soft-error has been a major concern for soft-errors in memories until late 1980s. Threat from environmental neutron-induced soft-error is growing with a rapid pace from early 1990s as device scaling proceeds. General features in charge collection mode soft-errors are reviewed and analyzed based on simulation results from the Monte Carlo soft-error simulator CORIMS. Combining with conventional charge collection mechanisms causing soft-error in SRAMs, new bipolar error mechanisms are confirmed under high-energy neutron-accelerated tests. The error mode is initially referred as "battery effect" and then referred as multi-coupled bipolar interaction (MCBI), based on theoretical/simulation studies with 2- or 4-bit 3D SRAM models. Countermeasure combining error correction code (ECC) and interleave turned out to be robust against this novel error mode.

6.1 Fundamentals of SER in SRAM Cell

When an energetic ionizing particle penetrates into a semiconductor device and pass through a depletion layer under a storage node (or diffusion layer), a single event effect (SEE) may take place.

As electronic devices are exposed to the harsh radiation environment from galactic cosmic rays or solar flares, a wide variety of SEEs in semiconductor devices caused by protons or heavy ions have been investigated for space applications over many years [6.1]–[6.5]. SEEs due to terrestrial neutron have been also investigated in avionics application [6.6, 6.7]. Although SEEs at the ground due to both alpha ray and terrestrial neutron were pointed out in the late 1970s [6.8, 6.9], major researches and countermeasures have focused almost solely on alpha ray-induced single event upset (SEU; soft-error) in memory devices until 1980s [6.10]. Details of the fundamentals of alpha ray-induced SEU are introduced in Sect. 6.2.

Terrestrial neutrons have been recognized as looming source of SEUs since early 1990s and concerns are now growing to wider and deeper extent than ever overwhelming alpha ray soft-error concerns [6.11]–[6.16].

As device scaling proceeds, the charge collected to a storage node due to alpha particle decreases because, in addition to shrinkage of charge collection volume, the charge density along an alpha particle path is only about 5–10 fC/μm, whereas the charge density created by heavier secondary ions (C,N,O,...,Ta,W) produced from nuclear spallation reaction between a high-energy neutron and a nucleus in the device is as high as 100–200 fC/μm [6.17]. This is why neutron-induced SEUs are getting dominant as the device size shrinks, compared to the contribution of alpha particles.

Table 6.1 Various modes in environmental-radiation-induced error

Category	Mode		Memo
Soft-error (SEU)	SBU (single bit upset)/SBE (single bit error)		Single bit upset/error in a memory device
	MCU (multi-cell upset)/MCE (multi-cell error)		Multiple-bit upset/error in one event
	MBU (multi-bit upset)/MBE (multi-bit error)		MCU/MCE in the same word (not correctable by ECC)
	Block error or k-cell error		Multiple bit errors along with BL or WL, originally due to errors in peripheral circuits
	MCBI (multi-coupled bipolar interaction)		Multiple-bit upset due to parasitic bipolar action triggered by snapback in channel. Correctable by rewriting. Sometimes associated with low currents. This is also called as battery effect formerly
	FBE (floating body effect)		Main error mode of SOI. Mitigated by body-tie
	SET (single event transient)		Transient noise in logic devices such as latch, inverter, and clock
	MNT (multi-node transient)		Multiple node SETs in a sequential/combinational logic circuit
Pseudo hard error	SESB (single event snapback)		Bipolar action in S/D channel. Impact-ionization may affect
	PCSE (power cycle soft-error)	SEL (single event latchup)	High current continues to flow due to parasitic thyristor. Only power cycle can resume, but sometimes destructive (hard error)
		SEFI (single event functional interrupt)	PCSE of logic devices or error corrected by resetting flip-flop states
		Firm error	Error mode of SRAM-based FPGA
Hard error	SEGR (single event gate rupture)		Destruction of thin oxide layer mainly due to high-energy heavy ion. Possible error mode in power MOSFET.
	SEB (single event burnout)		Destructive/explosive error of power MOSFET

A number of new modes of SEEs including both soft- and hard errors are evolving as device scaling proceeds as summarized in Table 6.1. SBU represents single bit upset or single bit error, and MCU represents multi-cell upset or multi-cell error. MCU is an error mode by which multiple memory cells are corrupted in one particle incident. MBU (multi-bit upset or multi-bit error) is MCU in the same word, which cannot be corrected by ECC. Block error or k(kilo)-cell error is specific to DRAM peripheral circuits [6.18]. MCBI represents multi-coupled bipolar interaction or battery effect. This mode is explained in more detail in Sect. 6.4. Floating body effect (FBE) is based on bipolar amplification in SOI devices [6.19]. Single event transient

(SET) and multi-node-transient (MNT) are relative to radiation-induced pulse in logic circuits [6.20]–[6.25]. All these error modes are included in soft-error (SEU).

Pseudo-hard error can cause either soft-error or hard error depending on incidental conditions. SESB stands for single event snapback [6.26]. PCSE represents power cycle soft-error and cannot be corrected by rewriting but can be recovered by power-cycling. PCSE may include single event latchup (SEL) [6.27, 6.28]. single event functional interrupt (SEFI) cannot be corrected by rewriting but can be corrected by resetting flip-flops (F/Fs) to default state [6.29]. Firm error is defined for SRAM-based field programmable gate array (FPGA) [6.30].

Hard error includes single event gate rupture (SEGR) [6.31] and single event burnout (SEB) [6.32, 6.33] and may take place in power devices.

The critical level of SEU rate was initially fixed at 1,000 FIT/chip (1FIT=1 error in 10^9 h) regardless of the type and density of the devices [6.34]. This requirement is almost equivalent to a rate as low as 1 error in 114 years, but it can also be equivalent to a rate of 1 error in a month even for small server systems with only 1,000 chips. While the error rate of 1,000 FIT in multimedia like movies or music obviously does not matter, it is definitely not acceptable for safety-critical or mission-critical systems such as those found in automotive and emergency rescue systems. In addition, it is becoming very difficult to keep the 1,000 FIT/chip level as devices continue to be scaled down and highly densified. The critical level of error rate, therefore, may be determined according to the error mode as summarized in Table 6.2. Very low target level may be given for fatal errors such as MBU, SEFI, SEL, and hard errors.

Table 6.2 Requirements for terrestrial neutron-induced error rate

Category	Error mode	Possible requirement	Remarks
Soft-error	SEU (single event upset)	1,000 FIT/chip (ITRS)	MTBF* $\sim$ 100 years/chip, <1 day/supercomputer *Mean time between failures
		200–400 FIT/Mbit or higher	• Can be corrected by ECC • Depend on application
	MCU (multi-cell upset/error)	A few to 10% of total SEUs or higher	• Fatal only if multiple bits are in the same word • Depend on application
	SET (single event transient)	To be determined	• Faulty noise in logic devices
	MBU (multi-bit upset)	0–10 FIT/chip	• MCU in the same word • May cause system crash
Pseudo-hard error	SEFI (single event functional interrupt)	0–1 FIT/chip or system	• Error of logic system • May be recovered by reset
	SEL (single event latchup)	0–1 FIT/chip or system	• Continuous current flow • Recovered by power cycle
Hard error	SEB (single event burnout), SEGR (single event gate rupture)	0–1 FIT/chip or system	• Permanent • May be fixed by re-mapping

Since even the target rate 1000 FIT is extremely low as mentioned before, it is very difficult to quantify the susceptibility of a particular device with any simple methods. Accelerated test methods and numerical simulation techniques, therefore, are indispensable for classifying and clarifying mechanisms, and quantification of current and future trends in SEEs. As for neutron-induced SEU, reliable solutions with (quasi-) monoenergetic and spallation neutron tests [6.17] and numerical simulation package cosmic ray impact simulator (CORIMS) have been developed within the framework of SECIS (SElf-Consistent Integrated System for neutron-induced soft-error) [6.16, 6.35]–[6.37]. The details of experimental methods are summarized elsewhere with relevant testing standards [6.38].

Particular attention has been paid on terrestrial neutron-induced SEUs in SRAM devices since SRAMs are being recognized as the most vulnerable device as device scaling proceeds.

The typical trends for SRAM and DRAM in literature data are summarized in Fig. 6.1 [6.15]. Trends in SEU cross section [= number of errors/particle fluence (n/cm^2)] of SRAM and DRAM are totally different: SRAM is getting worse as the memory density increases, while DRAM is getting better. The origin of these conflicting trends and further scaling trend will be elucidated in Sect. 6.3 with the help of CORIMS. New type of MCUs is evolving from around 130 nm SRAMs. The MCU mode is widely investigated and identified as a bipolar amplification effects in p-well by accelerated neutron irradiation methods and device/circuit simulations as described in Sect. 6.4. MBU (MCUs in the same word) is much more crucial than SBU for practical applications since MBU cannot be corrected by the standard error correction code (ECC) and thus cause occasionally costly unexpected shutdown of large-scale server systems [6.39]. Unlike alpha particles,

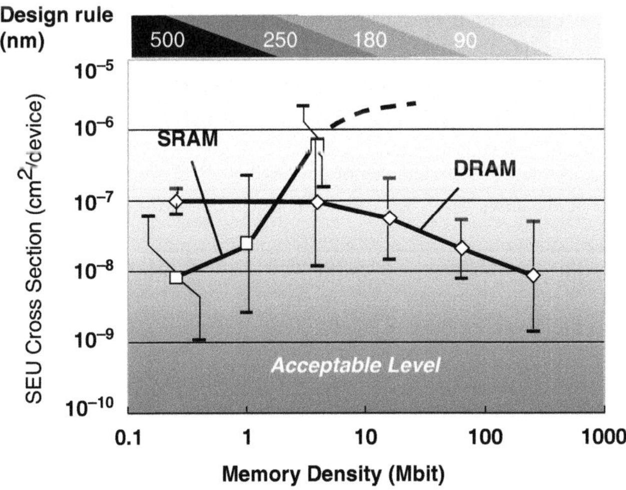

Fig. 6.1 Trends of soft-error susceptibility in SRAM and DRAM as a function of memory size. SEU cross section is defined as the number of errors divided by fluence (nucleons$/cm^2$)

heavy secondary ions evolved by nuclear spallation reaction are believed to spark MCUs [6.40]. Since such ions have relatively short range of typically less than 30 μm in silicon, interleave techniques, where bits in the same word are set apart with interval distance sometimes as far as 40–50 μm, are applied to suppress MBUs. Such techniques, however, would make architecture design sophisticated or even adverse performance (speed and power dissipation). More feasible scheme of ECC is presented in Sect. 6.5.

6.2 SER Caused by Alpha Particle

This section describes the fundamentals of the alpha particle soft-error. The alpha particle is emitted from the package materials, e.g., Pb solder bumps and molding compounds in semiconductor devices. As summarized in Table 6.3 with relevant isotopes and their characteristics, ^{210}Po, ^{232}Th, ^{238}U, and ^{235}U are major contributing radioisotopes. Since these materials are difficult to obtain (^{238}U and ^{235}U are nuclear fuel materials and are not commercially available), ^{241}Am foil is, instead, commonly used to quantify the susceptibility of semiconductor devices to alpha ray soft-error. If alpha particle hits the diffusion layer which corresponds to one node of a cell, the charge is generated and collected into the layer. As a result, the node voltage is changed. The critical charge Q_{crit} is defined as the minimum amount of charge to flip the data stored in a memory element (memory and latch/flip-flop) and one of the most important metrics in soft-error analysis and tolerance design. If the collected charge is large enough compared to the critical charge Q_{crit}, the memory element loses its data.

Early SRAM devices, contrast to DRAM devices, were more robust against the soft-errors because of a bi-stable circuit configuration [6.41]. Alpha particle-induced soft-errors [6.42]–[6.44] were a major obstacle to reducing cell sizes and increasing the scale of integration particularly for DRAM. Overcoming these

Table 6.3 Alpha ray source in semiconductor devices

Radio-isotope	Radioactive part	Decay constant (decay/s)	Energy of alpha ray and its emission ratio (%)
^{210}Po	Pb solder bump	5.794×10^{-8}	5.304 (100)
^{232}Th	Molding compound	1.568×10^{-18}	3.83 (0.2), 3.952 (23), 4.01 (77)
^{238}U	Molding compound	4.915×10^{-18}	4.039 (0.23), 4.147 (23), 4.196 (77)
^{235}U	Molding compound	3.122×10^{-17}	4.1525 (0.9), 4.2157 (5.7), 4.3237 (4.6), 4.3641 (11), 4.37 (6), 4.3952 (55), 4.4144 (2.1), 4.5025 (1.7), 4.5558 (4.2), 4.597 (5)
^{241}Am	Not contained in devices. Used only for test, normally in thin foil	5.096×10^{-11}	5.2443 (0.002), 5.3221 (0.015), 5.3884 (1.4), 5.4431 (12.8), 5.4857 (85.2), 5.5516 (0.2), 5.5442 (0.34)

soft-error problems has required the development of new device technologies, i.e., structures, fabrication processes, and materials, as well as new circuit technologies. The corrugated structures of stacked capacitors [6.45] and trench capacitors [6.46] have been useful in that they increase the electrode area and thus the magnitude of capacitance per unit area at the device surface. Increasing the capacitance per unit volume of capacitors by forming their dielectric layers of high-ε materials, e.g., Ta_2O_5 [6.47, 6.48] and $BaxSr_1$-$xTiO_3$ (BST) [6.49, 6.50] has also been useful. These and other techniques have alleviated the effect of the decreasing numbers of stored electrons with the decreasing sizes of DRAM cells.

Recently, SRAM cells fabricated with advanced CMOS technology have been subject to increased soft-error rates (SERs) [6.15, 6.51]–[6.54]. As a function of technology node, the SER has been studied by extrapolating the results of alpha particle experiment, and it has been reported that the real time SER (RTSER: Actual SER in the field) [6.38] increases with each successive node [6.41]. This is because the parasitic capacitance of SRAM cells decreases as technology advances. For LSIs of the next and subsequent generations, soft-errors in SRAMs are estimated to present a more serious problem than those in DRAMs as described in Fig. 6.1.

Here, the fundamental analysis of the alpha particle soft-error is described for SRAM cell. When the n^+ or p^+ storage node is hit by a charged particle including alpha particle, electron-hole pairs are generated along the ion path, as is schematically illustrated in Fig. 6.2, and they move along with the electric field in the depletion layer around the storage node, distorting the electric field, and thus elongate the electric field. As a result, the amount of the collected charge is more than the charge generated in the initial depletion layer. This effect is called "funneling" and is the origin of a soft-error. The collected charge depends on alpha particle incident position, angle, and energy. Three-dimensional (3D) device simulation [6.55] has been used to analyze this phenomenon. The injection of an alpha particle into silicon devices generates electrons and holes, which deform the potential distribution. The funneling length [6.56] was derived as the measure of effective length of carrier collection through the analysis of such potential deformation.

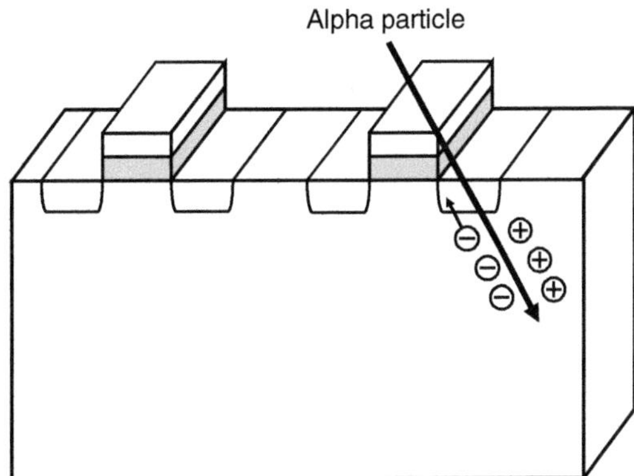

Fig. 6.2 Schematic of charge generation caused by alpha particle

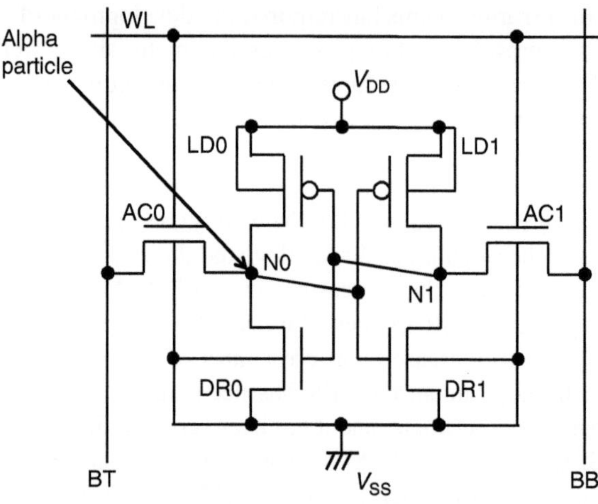

Fig. 6.3 An alpha particle hit to node N0 in cell

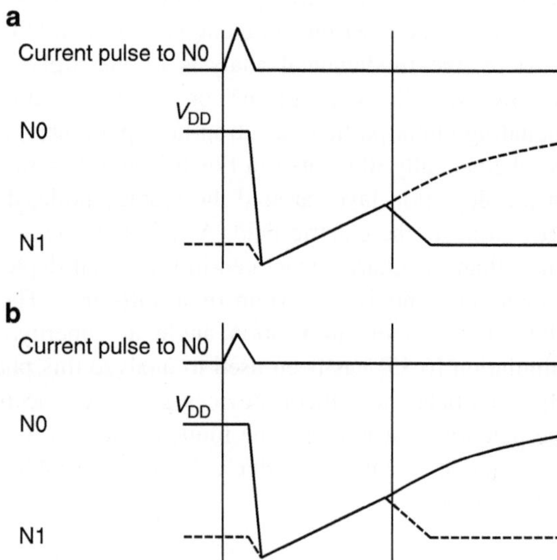

Fig. 6.4 Behavior of cell nodes following an alpha particle hit; (**a**) occurrence of soft-error, and (**b**) recovery to correct data

SRAM cell has a bi-stable circuit configuration. Therefore, inverter-level circuit simulation is necessary for analyzing soft-errors in SRAM devices. The circuit is shown in Fig. 6.3, in the case where an alpha particle hits the drain junction (N0) of the off-state MOSFET (DR0). The node N0 is at high voltage. Figure 6.4 shows the circuit simulation results. The current pulse is input to the node N0 as the results of the alpha particle hits. The magnitude of the current pulse is proportional to the total charge injected into the diffusion layer. After the current pulse input, the voltage level of N0 decreases. The voltage at the opposite node (N1) was pulled down synchronously due to the existence of the coupling capacitance between nodes N0 and N1. After the current pulse input, the levels of N0 and N1 increase simultaneously. If the total charge injected into the N0 is larger than Q_{crit}, the cell flips (Fig. 6.4a). If the total charge is less than Q_{crit}, the cell recovers to correct data (Fig. 6.4b). Q_{crit} has a proportional dependence on V_{DD} and can be increased by adding capacitance to the cell [6.57]. The SER is estimated by estimating the probability that the charge of more than Q_{crit} is collected to a cell node by alpha particle incident [6.58]. Experimental and theoretical study [6.59]–[6.64] of the physical mechanisms responsible for the soft-errors has yielded design guidelines for high-soft-error tolerance.

It is noteworthy to point out that the contribution of alpha ray to soft-error in SRAM devices is resuming again from late 2000s to the same level of that of neutrons due to significant decrease in Q_{crit} [6.65, 6.66]. The impact, however, may be limited since most alpha ray-induced soft-errors cause only SBU.

6.3   SER Caused by Neutrons and Its Quantification

6.3.1   Basic Knowledge of Terrestrial Neutrons

High-energy neutrons, protons, pions, muons and neutrinos are primarily produced by nuclear spallation reactions of extremely high-energy cosmic rays (mainly protons) with atmospheric nuclei (nitrogen and oxygen) as illustrated in Fig. 6.5 [6.67]. Charged particles are halted in a relatively short range, but neutrons produce a cascade of spallation reactions (air shower) that eventually make terrestrial neutrons at the ground level. Since charged particles twine around magnetic force lines, the geomagnetic and heliomagnetic fields act as shields against low-energy cosmic rays. Air also acts as a shield against neutrons, so that neutron flux varies with the location on the earth and solar activity. The neutron energy spectrum at the sea level in New York City (NYC) is shown in Fig. 6.6 [6.68]. The terrestrial neutron flux at the sea level is about $20\,\mathrm{n/cm^2/h}$ ($E_n > 1\,\mathrm{MeV}$).

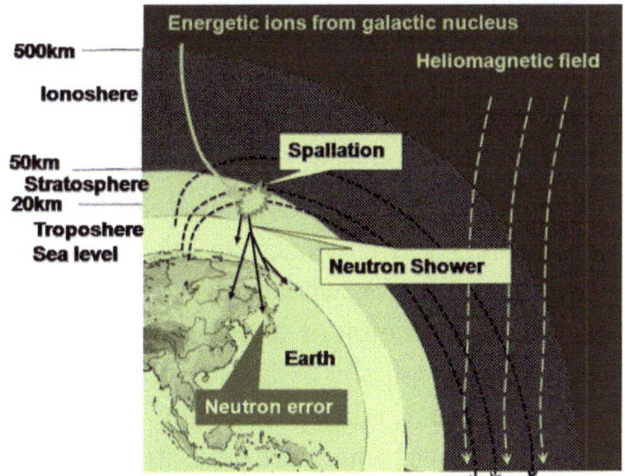

Fig. 6.5 Macroscopic mechanism of a neutron-induced soft-error (© 2002 IEEE)

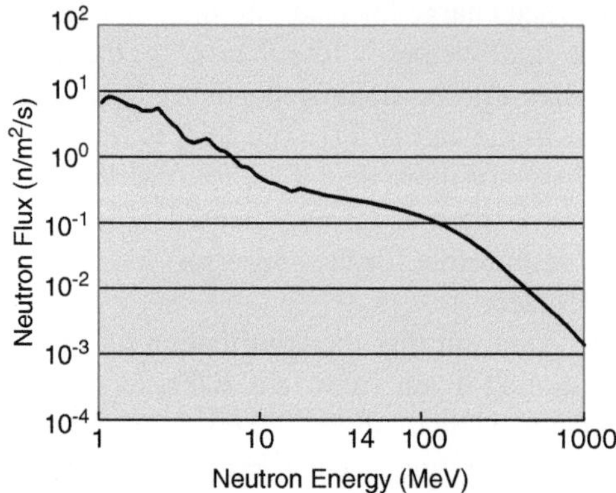

Fig. 6.6 Terrestrial neutron spectrum at NYC sea level (JESD89A)

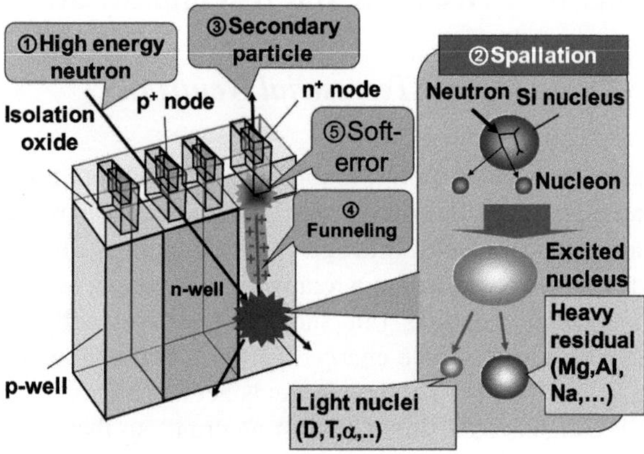

Fig. 6.7 CMOS SRAM structure and the microscopic mechanism of terrestrial neutron soft-error. Events proceed sequentially from (1) to (5)

A simplified bird's eye view of one bit CMOS-SRAM (static random access memory) cell is illustrated in Fig. 6.7 with a physical model of neutron-induced soft-error. The n-well (pMOSFET) is placed at the center of the SRAM device sandwiched by p-wells (nMOSFETs). The MOSFET channels are isolated by shallow trench isolation (STI). When a nucleus in the device undergoes a collision with a ballistic neutron, a nuclear spallation reaction, in which the nucleus breaks into secondary fragments, can take place with a certain probability. Similarly to alpha ray soft-error, when the storage node (diffusion layer) is hit by a secondary ion, a certain amount of electrons/holes produced along the ion track are collected to the nodes, typically by the funneling mechanism [6.69] and/or the drift-diffusion process. An SEU takes place when charge collected to the node exceeds the critical charge Q_{crit} over which the data "1(high)" in the node changes to "0(low)."

6.3.2 Overall System to Quantify SER–SECIS

The SER evaluation techniques using high-energy particle accelerators are integrated as an SER evaluation system, SECIS (SElf-Consistent Integrated System for SER evaluation system) [6.16, 6.36], combined with field testing and measurements of environmental factors. SECIS consists of five closely interlinked key techniques: (1) field testing of typical devices, (2) measurement of SEU cross section as a function of neutron energy (E_n) using mainly quasi-monoenergetic neutron beams [6.70]–[6.73] along with a necessary correction using the numerical simulation package CORIMS (developed for nuclear spallation and charge collection physics in the device [6.67, 6.74], (3) measurement of the terrestrial neutron spectrum at a specific location [6.75], (4) measuring geographic coordinates and terrestrial neutron dose in the field, and (5) a numerical simulation by CORIMS of field and accelerator testing of memory devices [6.16, 6.36]. The ultimate goal of this system is to evaluate SER of devices directly by the simulator CORIMS. It is expected that repetition of the procedure from (1) to (5) converges the evaluated value of SER obtained by SECIS with a high degree of accuracy. In order to confirm the usefulness of SECIS, a comparison among SER values obtained by field testing, accelerator testing, and simulation is carried out. Figure 6.8 demonstrates the series of the SER values of low-power consumption CMOS SRAM with 180 nm process technology at three different locations in Japan at altitudes of 86, 755, and 1,988 m with the simulation results from CORIMS [6.73].

6.3.3 Simulation Techniques to Quantify Neutron SER

6.3.3.1 Nuclear Reaction Model

The simulation packages CORIMS is equipped with vigorous numerical algorithms for nuclear spallation reactions, ion track analysis in an infinite cell matrix, and charge collection to the storage nodes. A simplified 3D device layout, using 3D

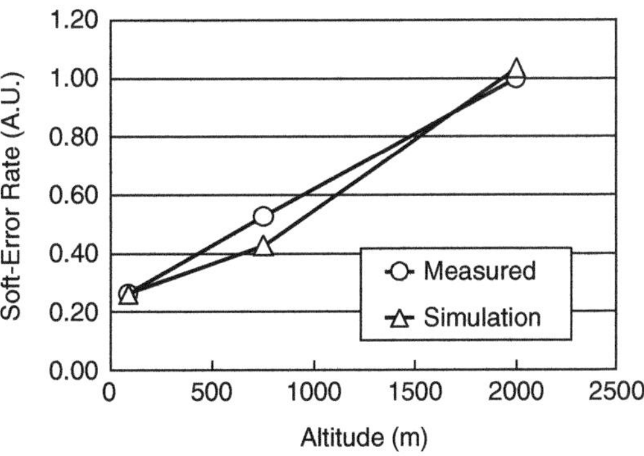

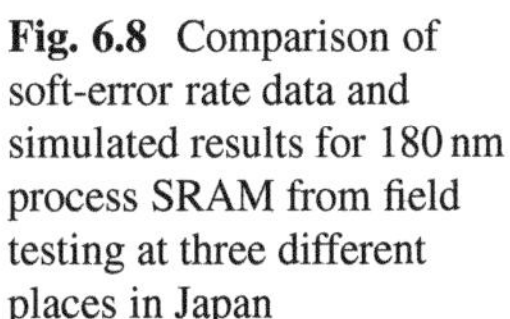
Fig. 6.8 Comparison of soft-error rate data and simulated results for 180 nm process SRAM from field testing at three different places in Japan

CAD parameters from GDS2 files, including storage nodes, channels, wells, and isolation oxide are automatically implemented in the device model with an effective and systematic ion track analyses.

The Monte Carlo method is applied to a number of steps in the physical process, from nuclear spallation reactions, to charge collection to storage node, or MCU/MBU analyses.

The SER of any MOSFET device is calculated in any given radiation environments, such as the terrestrial neutron field at any location on the earth and neutron fields at accelerator facilities with a given neutron spectrum.

Charge production from direct irradiation by ions can also be treated, for example, alpha source (^{258}U, ^{232}Th, ^{210}Po, etc.) contamination is implemented.

The intra-nuclear cascade (INC) model [6.76] is applied to the prompt collision process, where the many-body collision among nucleons (neutrons and protons) is treated numerically as a cascade of relativistic binary collisions between two nucleons in the nucleus.

The evaporation model [6.77] of light particles from an excited nucleus is also applied for the delayed nuclear reaction process, where nucleons (n and p), deuterons (^{2}H or D), tritons (^{3}H or T), helium, and residual nucleus are released. The type, energy, and direction of each secondary ion are automatically determined in the Monte Carlo nuclear reaction simulation.

6.3.3.2 The Single Bit Model

As the model layout of a MOSFET SRAM cell illustrated in Fig. 6.7, the active regions are isolated by an STI oxide in the lateral direction. In addition, the wells line up across the word lines, so charge collection in the lateral direction is tightly limited in the present SRAM device. This boils down to that the possibility of MBU is essentially low in this device, because bits in a word are aligned along the word line (lateral direction).

The charge collection mechanisms (funneling and drift-diffusion mechanisms) are implemented into CORIMS.

The secondary ion does not necessarily pass through the depletion layer in the drift-diffusion mechanism. The hole and electrons drift or diffuse into the storage node. In CORIMS, this effect is implemented simply by a drift-diffusion "box" beneath the storage node.

6.3.3.3 The MCU Model

As is introduced later in Section 6.4, the characteristics of MCUs are found to depend significantly on the data patterns CHB, CHBc (complement of CHB or inverse pattern of CHB), All0, and All1 [6.76]–[6.78]. The layout of bits in a word is of primary importance in designing the ECC and interleaving parameters. In the cell matrix model of CORIMS, these parameters can be implemented and

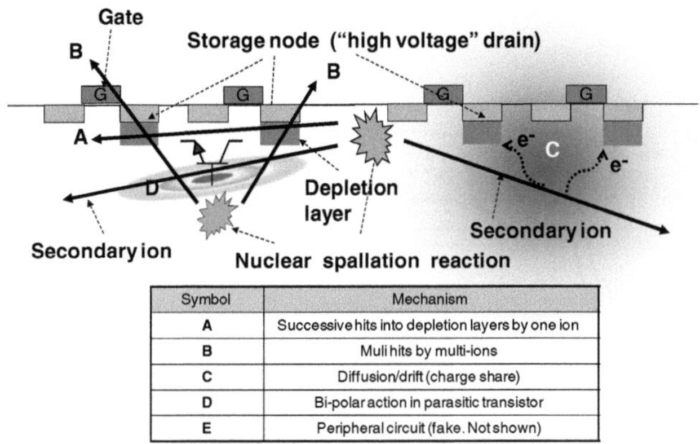

Symbol	Mechanism
A	Successive hits into depletion layers by one ion
B	Muli hits by multi-ions
C	Diffusion/drift (charge share)
D	Bi-polar action in parasitic transistor
E	Peripheral circuit (fake. Not shown)

Fig. 6.9 Five possible mechanisms of MCU

their sensitivity can be surveyed in the infinite number of memory cell matrix. As summarized in Fig. 6.9, there are five possible MCU mechanisms: (a) a successive hit by one secondary ion; (b) multi-hits by multi-ions; (c) charge sharing by adjacent nodes; (d) parasitic bipolar action between a S/D channel; and (e) fake MCU caused by failure in the peripheral circuits (e.g., decoder/encoder. This mode results in SEFI). The bipolar modes including (d) and the peripheral circuit mode (e) in Fig. 6.9 are very crucial modes, but they are not modeled in the CORIMS currently because of their intrinsic difficulty in simple modeling. Together with pseudo/full 3D device simulations, the in-depth study will be introduced in Sect. 6.4.

6.3.3.4 Validations of the Simulation Technique

SER simulation models in CORIMS have been validated in a number of phases/situations through nuclear reactions, accelerator testing, and testing results in the field.

For validation of the nuclear reaction model, nuclear reaction data of protons on ^{27}Al [6.78], as well as some other reactions, are used as demonstrated in Tang's work [6.79]. The agreements between literature data and calculation results seem acceptable.

For validation of the combined models for nuclear reactions and charge collection/diffusion to the storage, results from accelerator-based testing are used. The simulated results are shown in Fig. 6.10 for 4, 8, 16 Mbit 130 and 180 nm SRAMs using (quasi-) monoenergetic neutron tests [6.71]. The errors between the simulations and the data are around 20–30%, in average.

As demonstrated in Fig. 6.8 as a function of altitude, the simulated and measured results in the field tests conducted in three locations in Japan show good agreement with each other within ±20% error.

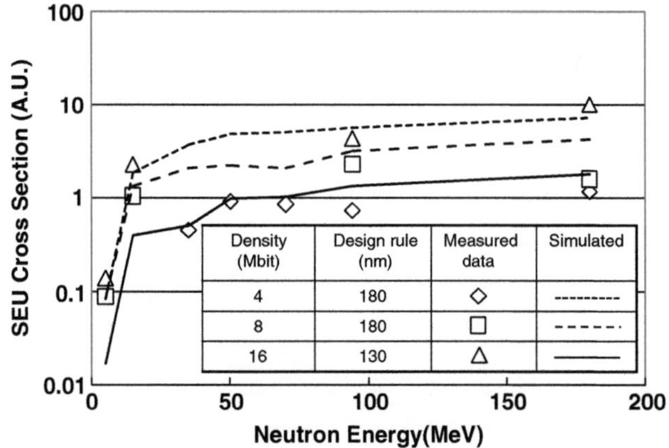

Fig. 6.10 Comparison between experimental and simulation results from CORIMS for (quasi-) mono energy test results

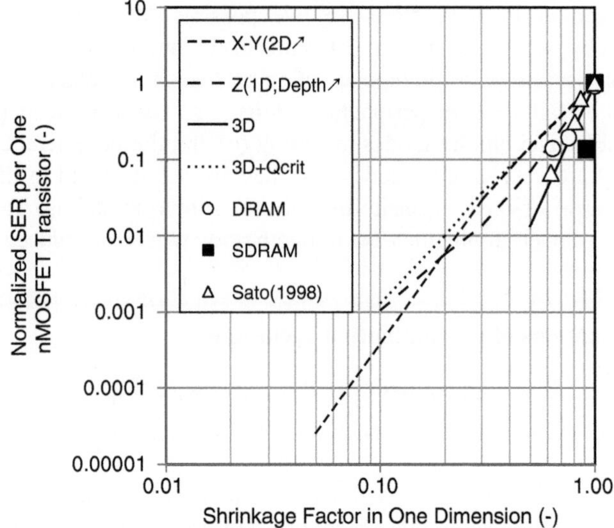

Fig. 6.11 Effect of scaling on soft-error rate (*white circle, black square*: measured at LANSCE by Eto [6.11]; *white triangle*: Sato et al. [6.80]; k is calculated from the square root of the dimensionless area of the device). Lines are simulated results from CORIMS [6.16] (© 2001 IEEE)

6.3.4 Predictions of Scaling Effects from CORIMS

Figure 6.11 shows simulated results from the effects of spatial scaling and reduction of critical charge on the SEU cross section per bit of DRAM [6.16], together with experimental results measured at LANSCE [6.80, 6.11]. It can be said that spatial scaling has a significantly beneficial impact on SEU susceptibility: 1D spatial scaling with a factor k results in roughly k^2 scaling of SEU susceptibility. As for 3D scaling, k^6 scaling of SEU susceptibility is expected. Physically, this can be interpreted that the SEU susceptibility is in proportion to the product of two volumes (6D in total) in the device namely, the storage node volume which determines the probability to be hit by the secondary ions, and the surrounding volume which determines the amount of charge collected to the storage node. It is also shown that if the critical charge Q_{crit} is scaled simultaneously, the beneficial effect of spatial scaling deteriorates significantly, as exemplified in the 3D and Q_{crit} case.

These simulated results give a clear explanation for the conflicting trends observed in Fig. 6.1: For DRAMs, where the critical charge has been strenuously kept high by engineering of the capacitance, the beneficial effect of spatial scaling has prevailed. By contrast, the critical charges of SRAMs are generally reduced with scaling due to reduction in parasitic capacitance. As a result, the beneficial effect is eventually blocked. Further decrease in the critical charge is an inevitable direction in the road map of ULSI devices as a consequence of reducing the supply voltage, area of the (parasitic) capacitors, and dielectric constant of the interlayer dielectrics so that increase in SEU susceptibility of memory devices is also inevitable in general.

It is well known that increasing critical charge is a very effective option to reduce alpha ray-induced SER as exemplified in Fig. 6.12 [6.81]. This is not true for terrestrial neutron-induced soft-error as also shown in Fig. 6.11. Figure 6.13 shows simulation results for total collected charge in the 130 nm SRAM [6.82]. Contribution of heavy ion (atomic number Z > 10), alpha, and protons produced through spallation reaction with a nucleus in the device to errors are also shown. It is noted that the heavy ion dominant region (collected charge is larger than about 10 fC) has dull response to the collected charge, which is consistent with the trend in Fig. 6.11 for terrestrial neutrons, while it is also noted that the proton (collected charge is less than about 2 fC) and alpha (collected charge is around 2–10 fC)

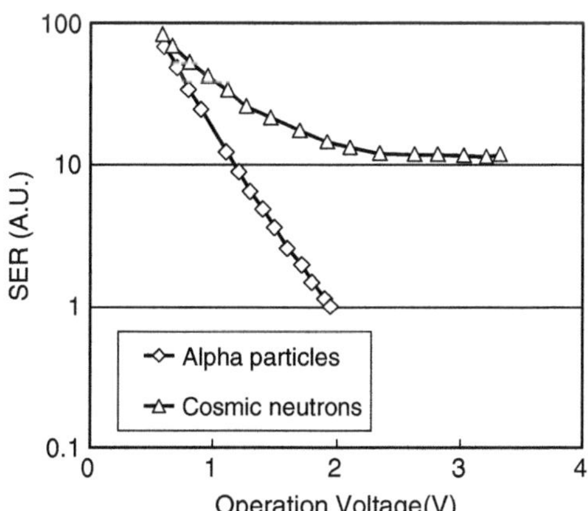

Fig. 6.12 Total collected charge to storage node (130 nm SRAM). Contribution of proton becomes higher at the lower total collected charge [6.41]

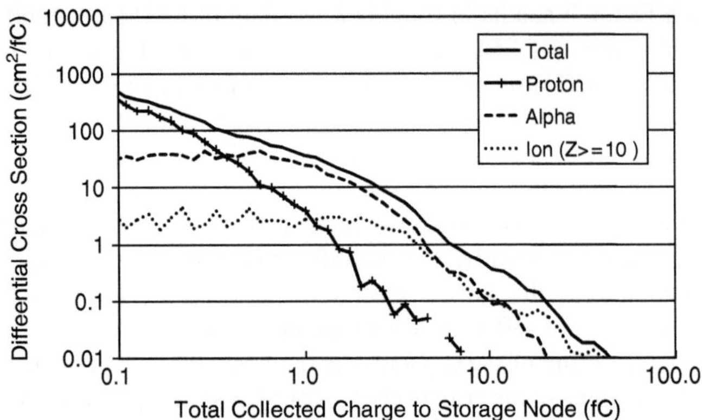

Fig. 6.13 Total collected charge to storage node (130 nm SRAM). Contribution of proton becomes higher at the lower total collected charge [6.82]] (© 2010 IEEE)

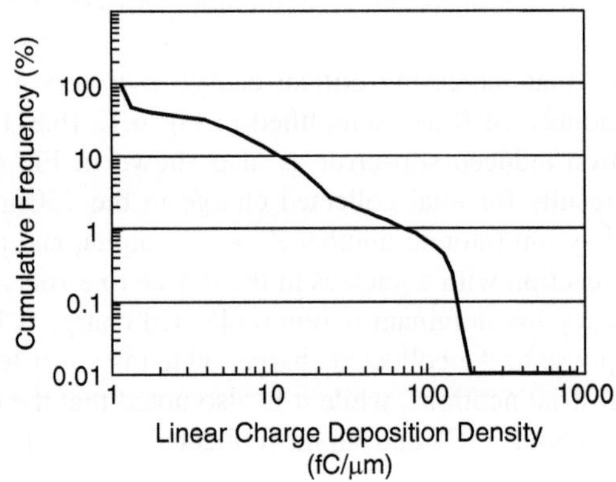

Fig. 6.14 Cumulative frequency of linear charge deposition density for terrestrial neutron spectrum in Tokyo [6.82] (© 2010 IEEE)

dominant regions have steeper trends than the heavier ion dominant region. This implies that reducing Q_{crit} may result in catastrophic increase in SER.

It is noteworthy that the absolute values in the curves, for instance the maximum collected charge, are subject to change with changing the device architectures.

Figure 6.14 may be useful for more general purpose. The figure shows the cumulative frequency of the charge deposition density in a Si substrate at the boundary of the sensitive volume [6.82]. As long as the material of a device is made of Si, the curve holds true under a terrestrial neutron environment. As a result, Fig. 6.14 can be used for the device-hardening design. For instance, if a device is tolerant to a charge density of 170 fC/μm, the device is absolutely immune to terrestrial neutron soft-errors.

6.4 Evolution of MCU Problems and Clarification of the Mechanism

An on-chip ECC circuit [6.83, 6.84] is helpful as a way to achieve a low SER. However, the ECC circuit is only capable of correcting one error at each address. That is, a multi-cell error where all of the affected cells belong to the same address are not corrected by the ECC circuit. However, the circuit does correct a multi-cell error where the affected cells all belong to different addresses. To increase the rate of multi-cell error correction by ECC circuits, we need to arrange the cell addresses, so that simultaneous cosmic ray-induced multiple errors are most likely to occur at different physical addresses. The first step in achieving this is to get a better understanding of the process of multi-cell errors.

This session describes the mechanism responsible for multi-cell errors. Through device- and circuit-level simulation, it is presented that a parasitic bipolar effect is responsible and found that the underlying mechanism is what we call a battery effect or parasitic thyristor effects triggered by a single event snapback in p-well (MCBI).

6.4.1 MCU Characterization by Accelerator-Based Experiments

6.4.1.1 DUTs and Neutron Beams

For the irradiation test, 130 nm SRAMs are used. The layout and structure are schematically shown in Fig. 6.7.

The test was carried out in Theodore Svedberg Laboratory (TSL) of Uppsala University [6.73] with neutron peak energies E_p of 21, 47, 70, 96 and 176 MeV. Specific automatic data analysis sequences are applied in space and time domains [6.85] since specific MCU mode as described in this section is found to be looming from 130 nm SRAM generation [6.86].

6.4.1.2 MCU Patterns

In total, 2,564 MCUs were identified without any MBUs. The spatial failed bit patterns along with bitlines (BLs) and wordlines (WLs) were classified into two groups of their initial data pattern, namely CHB/CHBc (complement of CHB) and All "0"/All "1". As described in the inset (see Fig. 6.7 for more details) in Fig. 6.15, it is noteworthy that the difference between the "0" and "1" data bits is only "high" node position in a SRAM bit. For "1" bit, "high" nodes are located in one side (right-hand side in the present case). For "0" bit, it is reversed. Based on topological analysis of MCUs as partly exemplified in Fig. 6.15, the following implications are also obtained [6.87, 6.88]:

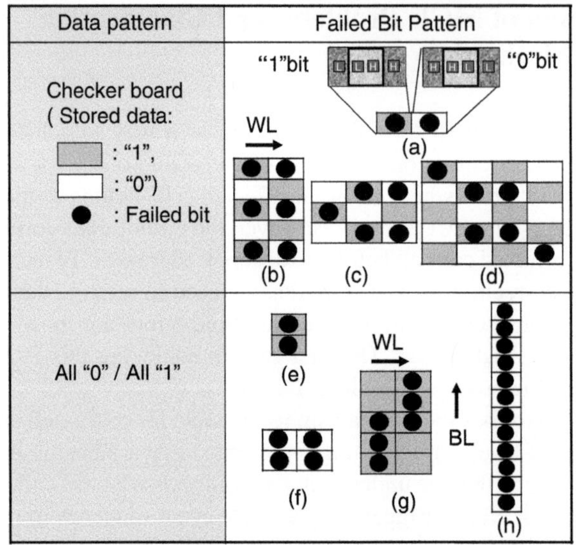

Fig. 6.15 Typical failure bit map of MCUs depending on the stored data pattern (© 2006 IEEE)

1. For CHB/CHBc, MCU with two error bits only aligns adjacently along with a wordline (WL) as seen in (a) and (b). No MCU with three error bits aligned along with a WL was found.
2. As rare cases for CHB and CHBc, the clusters cover multi BLs and WLs [(c) and (d)]
3. For ALL0/ALL1, MCU normally makes a single successive straight line along BL as seen in (e) and (h), which imply that "high" nodes aligned in the same p-well are subject to fail. As many as 12 successive MCEs are observed at the maximum as in (h).
4. Likewise for CHB and CHBc, "high" nodes in p-well are subject to fail showing "leap-frog" cluster error bit pattern along a BL as in (b).

Figure 6.16 shows unnatural multiplicity observed in the test results. For CHB/CHBc, it is found that the number of double bit error exceeds those of single bit error when the neutron energy becomes higher, while the trends for ALL0/ALL1 seems rather normal. This trend has not been reported anywhere else.

6.4.1.3 Influence of Tap Locations

Figure 6.17 shows error population of (a) single bit upset (SBU) and (b) MCU errors along with BL with modified address (Mod 128) to see the effects of tap locations [6.78]. Clear dependency with 32-bit intervals appears only on MCUs along BL direction, where tap for bias is located with 32-bit interval. This implies that:

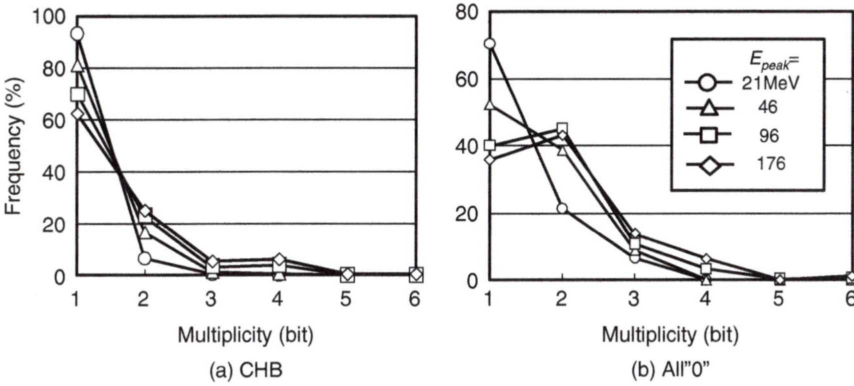

Fig. 6.16 Dependency of the MCU distribution on neutron peak-energy and data pattern for 130 nm SRAM obtained at TSL: (**a**) checkerboard pattern and (**b**) all "0" pattern [6.87, 6.88] (© 2006 IEEE)

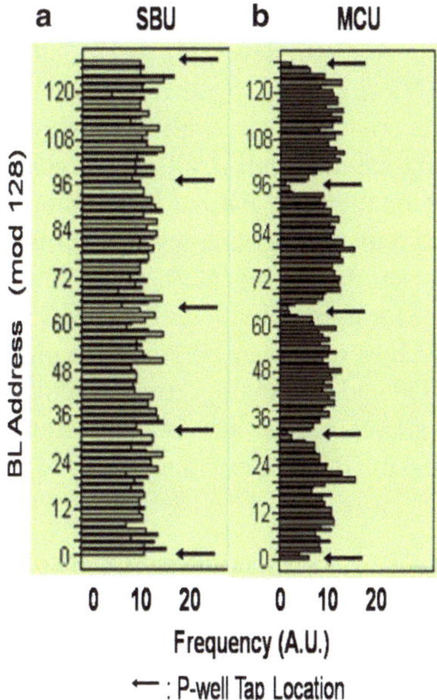

Fig. 6.17 Address dependency (mod 128) of single bit (**a**) and multi-cell upset (**b**) along BL [6.87] (© 2010 IEEE)

1. MCU probability is low at the 2–3 bit vicinity of the tap position, which implies that the resistance to the tap governs MCU.
2. Major mechanism of MCUs is different from SBUs. For SBUs, major SEU mechanism may be attributed to charge collection–diffusion or snapback mechanism, while that of MCUs must be related to *bipolar* action where resistance between parasitic transistors and taps plays major role.

The other evidence indicating bipolar action is stepwise distribution of I_{DD} increase. The number of discrete steps increases with the peak neutron energy and is believed to depend on the MCU bit-multiplicity [6.77].

6.4.2 Simplified 3D Device Simulation Mixed with Circuit Simulation

6.4.2.1 Combined Circuit and Device-Level Simulation to Model Multi-Cell Errors

The two-memory-cell (Cell_0 and Cell_1) model shown in Fig. 6.18 is used to simulate multi-cell errors at the circuit level. Nodes NR0 in INV_0 and NL1 in INV_1 are initially at the high voltage. Figure 6.19 shows our lithographically symmetrical cell layout [6.89, 6.90]. The p and n wells run parallel to the bit lines. The nMOS transistors in INV_0 and INV_1 are in the same p-well (P-well_1). Well taps are placed out of memory array. At circuit level (Fig. 6.18), substrate nodes of the nMOS transistors in INV_0 and INV_1 are tied together and connected to a parasitic p-well resistance (R_{well}), which is connected to a well tap. Using this model, we looked at the effect, on the adjoining memory cell (Cell_1), of incidence of a secondary ion generated by a cosmic ray on the node NR0 of the memory cell Cell_0.

We also modeled the circuit as a device-level structure on a three-dimensional device simulator; this is shown in Fig. 6.20. This simple structure is used to guarantee fast convergence for the calculations. The pMOS transistor in one inverter circuit (INV_0) is replaced by a resistance (R0) and the nMOS transistor is replaced by n^+ diffusion regions (D0 and S). Analogous replacements are made in the other circuit. S is a common source region and is connected to the ground. The p^+ diffusion layer is connected to GND to form a well tap. A parasitic p-well resistance

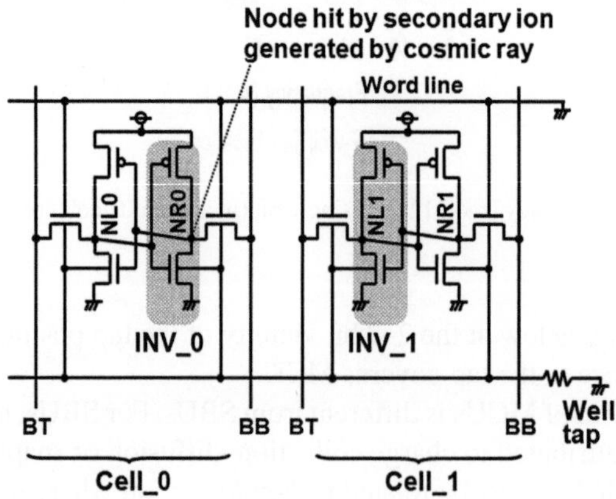

Fig. 6.18 The model used in circuit simulation

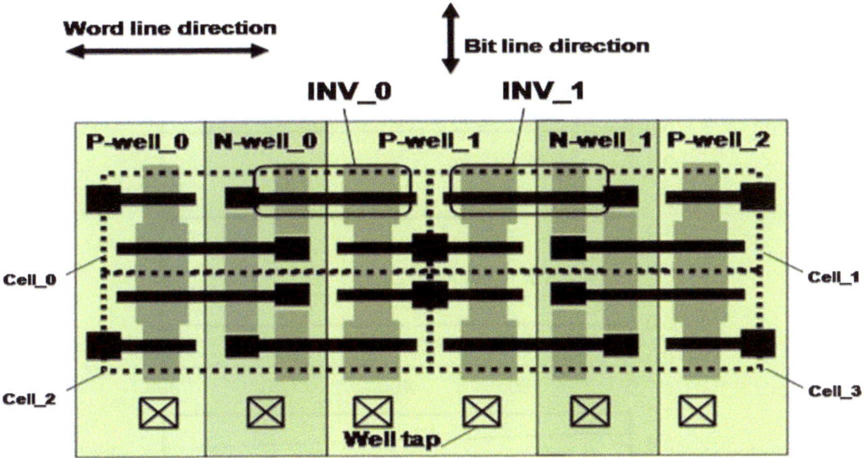

Fig. 6.19 The SRAM layout for circuit simulation. INV_0 and INV_1 are located in the same p-well

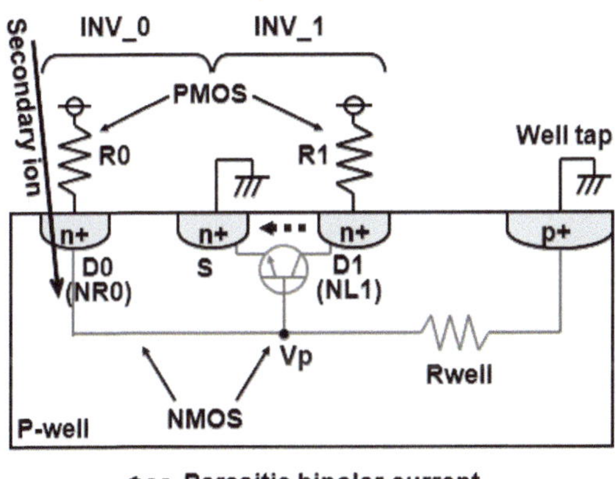

Fig. 6.20 The model used in device simulation

(R_{well}) is inserted between the nMOS transistors and the well tap. The parasitic bipolar device between D1 and S is included.

We examine the case where a secondary ion generated by a cosmic ray hits the diffusion layer (D0), which corresponds to one node (NR0) of Cell_0. At this stage, we ignore incidence in all other locations, because such events are believed to only negligibly contribute to the number of soft-errors [6.91]. We then use our three-dimensional device simulator CADDETH (Computer Aided Device DEsign in THree dimensions) [6.91] to calculate the voltage changes at D0 and D1 in response to the secondary ion hitting D0. CADDETH can be applied to direct current characteristic analyses for a wide variety of silicon devices. The simulation runs in the simulator's transient operating mode and involves solving Poisson's equation and the current-continuity equations.

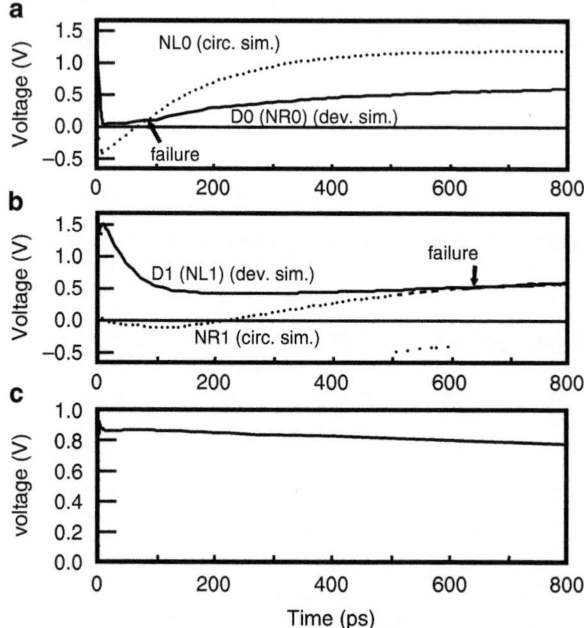

Fig. 6.21 Nodes and well voltages in device- and circuit-level simulation; (**a**) cell hit by the cosmic ray, Cell_0, (**b**) the adjoining cell, Cell_1, and (**c**) p-well voltage, i.e., voltage at V_p between D1 and S

The calculated voltage at D0 (NR0) is shown as the solid line of Fig. 6.21a. We then carried out circuit-level simulation of the SRAM cell using the circuits shown in Fig. 6.18, with the calculated D0 voltage waveform added as a voltage source to the voltage at NR0 in Cell_0. The resulting voltage on the opposite node (NL0) is shown as the dotted line of Fig. 6.20a. The relation between the levels on NL0 and NR0 is inverted within 100 ps of the cosmic ray strike. This constitutes a soft-error. We also used device-level calculation to obtain the voltage in response at D1 (NL1); this is shown as the solid line of Fig. 6.21b. We then performed circuit-level simulation of the SRAM, with the calculated D1 voltage waveform added as a voltage source to the voltage at NL1 in Cell_1. The resulting voltage on the opposite node (NR1) is shown as the dotted line in Fig. 6.4b. The relation between the levels on NL1 and NR1 is inverted within 700 ps of the cosmic ray strike. A further soft-error has thus occurred. Note that the voltage on NL1 decreases much more slowly than that on NR0.

6.4.2.2 Parasitic Bipolar Effect is Responsible for Multi-Cell Errors

The cell hit by the cosmic ray (Cell_0) loses its data through funneling. The adjoining cell (Cell_1) loses its data because of a parasitic bipolar effect. The mechanism is explained below, with reference to Fig. 6.19.

After the secondary ion hits NR0, funneling leads to the very rapid collection of electrons in NR0. This lasts for 10 ps and lowers the voltage on NR0 to zero. On the other hand, the holes generated by the ray remain in the p-well. The voltage in the p-well (V_p) thus floats up to 0.9 V, as shown in Fig. 6.21c. This floating p-well switches the parasitic bipolar element on which leads to a flow of current from NL1 to S. Thus, the voltage on NL1 slowly falls. This is the first description of how a parasitic bipolar effect leads to multi-cell errors.

A cosmic ray generates ten times as many carriers as an alpha particle [6.82]; so we repeated device-level simulation for various numbers of generated carriers. The number of carriers generated is reflected in the total charge, Q_0, generated in the depletion region struck by the ray. We thus use Q_0 as the index of the number of carriers. The results for p-well voltage, which switches the bipolar device on and off, are shown in Fig. 6.22. After the cosmic ray hits, the voltage in the p-well abruptly floats up to about 1.0 V. The maximum voltage in the p-well is determined by the built-in potential (the intrinsic barrier potential of the p–n junction). The p-well floating continues several nanoseconds. The floating time, which we define as the time the p-well voltage takes to cross back below 0.85 V, increases with Q_0. The explanation for this is as follows. A larger Q_0 reflects the generation of more electrons and holes in the p-well. This lengthens the periods over which the p-well supplies electrons to D0 and holes remain in the p-well. The floating time is correspondingly longer. However, the built-in potential means that the peak voltage never rises above 1.0 V and is largely independent of Q_0. We refer to this combination of phenomena as a battery effect; this is the effect which switches the parasitic bipolar device on. We use the term "battery" because the phenomena continue to supply the voltage of 1.0 V to the p-well like a function of a battery.

Now that we have identified the mechanism, let us summarize our investigation up to this point. We have shown, for the first time, how a parasitic bipolar effect leads to failure of the cells adjoining a cell penetrated by a cosmic ray. The parasitic bipolar element is switched on by what we term the battery effect.

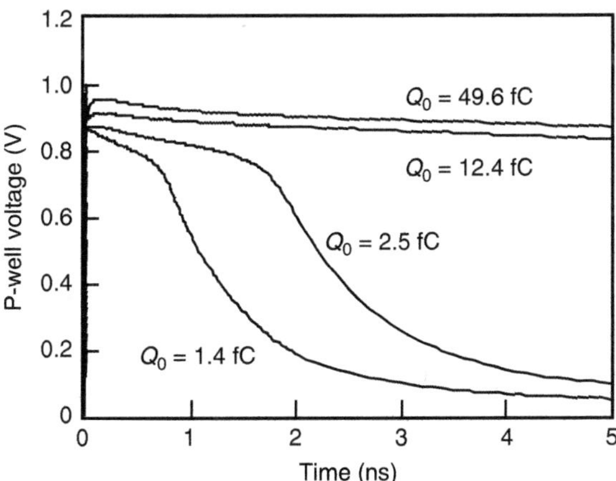

Fig. 6.22 Simulated p-well voltage after a cosmic ray hit for varying amounts of initial charge in the depletion layer, Q_0

6.4.3 *Full 3D Device Simulation with Four-Partial-Cell Model and Multi-Coupled Bipolar Interaction (MCBI)*

In the previous model, 2-bit pseudo 3D model is applied and bipolar amplification mechanism "battery effect" in the p-well is numerically identified. But the failed bit spatial patterns shown in Fig. 6.15 and associated I_{DD} stepwise current increase are not fully understood. The conventional charge collection model to a diffusion node is applied as the initial trigger mechanism in the model but its spatial impact may be limited because it is surrounded by STI walls. Instead, a breakthrough trigger mechanism and an entire bipolar mechanism are proposed in this subsection.

The 3D device simulator DAVINCI^{TM} is used for clear elucidation of newly unveiled neutron-induced MCU. The SRH [6.92] and Auger recombination [6.93] models, Lombardi's surface mobility model [6.94], and impact ionization model [6.95] are applied. The actual impurity profiles of 130 nm SRAM process are applied. The simulation results of $V_G - I_{DS}$ characteristics agreed satisfactorily with measured values.

1. Four partial SRAM memory cells are combined to make a 3D 4-bit model as shown in Fig. 6.23 where a full p-well structure is placed in the center to simulate a bipolar action therein. In Fig. 6.23, the poly-silicon gates are not

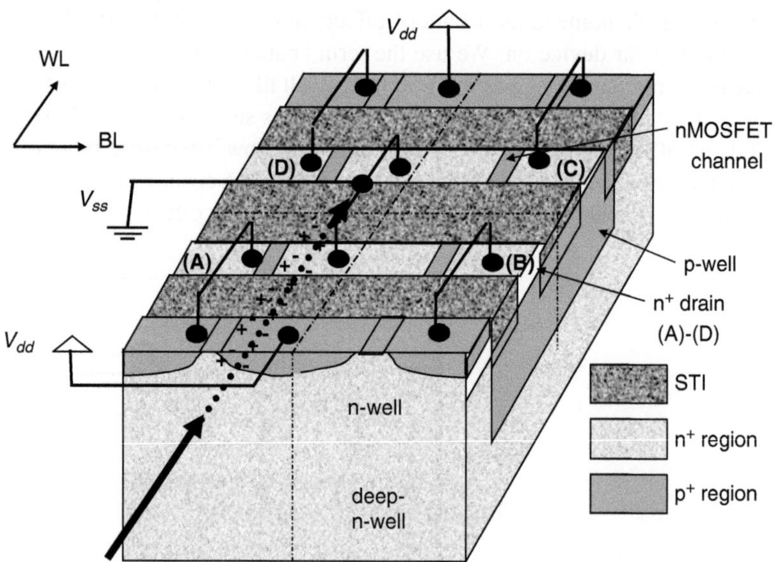

Fig. 6.23 Four partial bit SRAM model and Mg$^+$ ion bombardment across p–n junction in p-well [6.96]

explicitly shown for the simplicity, although they are actually implemented in the DAVINCI^{TM} model.

2. Magnesium ion of energy around 10 MeV (initial charge deposition density is 123 fC/μm) is chosen as a candidate secondary ion produced from spallation reaction of a silicon nucleus with terrestrial neutron. The Mg$^+$ ion bombarded from the n-well side of the four cell model, penetrating the p–n junctions in the side walls of the p-well as shown in Fig. 6.23. The Mg$^+$ ion runs below the drain nodes (A) and (D), and gets out of the p-well through the p–n junction in the other side of the p-well. Any p$^+$ or n$^+$ diffusion layers/nodes are not hit by the ion so that any charge collection due to funneling mechanism to the storage node cannot take place in this case In other words, soft-error cannot take place under this situation, based on the conventional soft-error mechanism.

Figure 6.24 shows typical simulation results for the currents through n$^+$ nodes in Nodes (A) to (D) in the p-well. Initially, the currents in Nodes (A) and (D) jump at the onset of Mg$^+$ ion due to single event Snap Back (SESB) mechanism. Then, the currents in Nodes (B) and (C) which are apart from the penetration points in the p-well increase gradually to the same level as in Nodes (A) and (D); eventually the current of a few milliampere continue to flow and cause soft-error in all four bits, which gives I_{DD} step current only slightly lower than the measured values and a quite similar scheme to the experimental observations. SESB is identified as a trigger mechanism of overall responses as summarized in Fig. 6.25. Namely, channel under the gate is turned on by SESB initially and the hole current flows to the ground. Then Tr 2 is turned on to the deep n-well. Finally, Tr 3 is turned on to make a parasitic thyristor circuit. Based on this mechanism, the potential in p-well is kept high. The storage nodes with "high" data in the p-well in the vicinity of the MCBI region are subject to be failed.

Figure 6.26 demonstrates that only MCBI can explain the MCU pattern dependency on data pattern: As for CHB/CHBc data pattern, a set of two "high" nodes

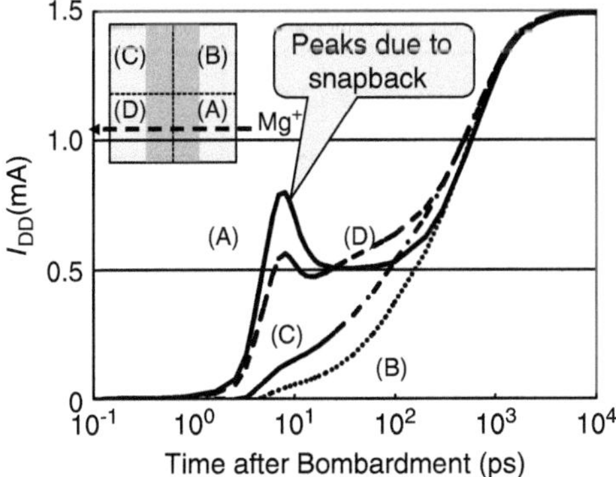

Fig. 6.24 Simulation results for drain current I_{DD} in Nodes (A)–(D) after 20 MeV Mg ion bombardment to result in MCU of 4-partial cells [6.96]

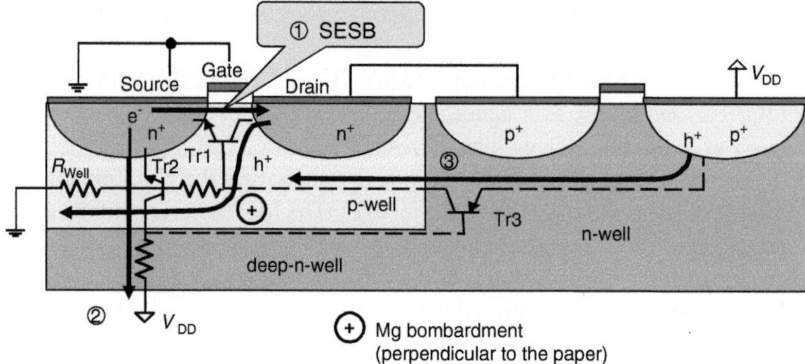

Fig. 6.25 Overall MCBI mechanism. Initially channel under the gate is turned on by single event snap back (SESB) and the hole current flows to the ground. Then Tr 2 is turned on to the deep n-well. Finally, Tr 3 (n-well layout is folded for the simplicity. See Fig. 6.23) is turned on to make a parasitic thyristor circuit [6.96]

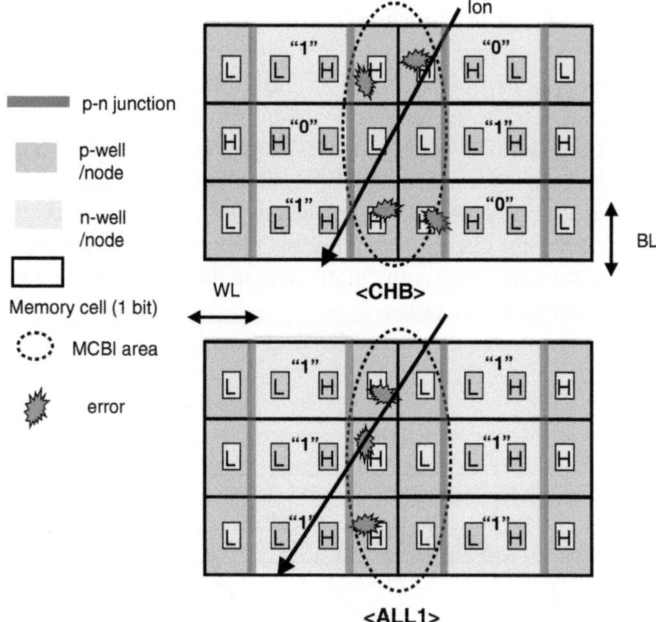

Fig. 6.26 Interpretation of dependency on error bit pattern on data pattern. Numbers in *quotes* denote data condition in each cell. "H" and "L" denote "High" and "Low" conditions in each storage node, respectively

aligns with one node interval along BL, so the leap-frog pattern appears as seen in Fig. 6.15b. Meanwhile, "high" nodes align in a single line without interval for ALL0/ALL1 data patterns; so a single successive error bit pattern appears as seen in Fig. 6.15h. In summary, the experimental findings and simulation studies strongly suggest a novel MCU mechanism, which we now call multi-coupled bipolar interaction (MCBI), is emerging.

As a model SRAM 130 nm process 6T CMOS SRAMs are irradiated by quasi-monoenergetic high energy neutrons. Distinctive differences in MCU features are observed between the data pattern groups CHB/CHBc and ALL0/ALL1, demonstrating the efficacy of automatic MCU classification technologies. Among diversity of MCUs, novel error propagation mechanisms in MCU, MCBI, is identified as a new neutron SEE mode of highly advanced semiconductor devices.

MCBI has five features: (1) the mode includes multicell upsets in a localized cell matrix; (2) accompanying nondestructive I_{DD} current increase depending on the number of multiplicity; (3) the mode is *not* power cycle soft-error (PCSE) so that it can be distinguished from neutron-induced latchup (typical PCSE mode), or more precisely; (4) it is caused by parasitic thyristor triggered by secondary ion-induced snapback; (5) the major mechanism of SBU is different from MCU. Major mechanism of SBU may be snapback only or contribution of light secondary particles such as proton and alpha particles.

Although the new MCBI mode turned out to be rewritable, neither destructive nor PCSE, it has potentially critical influences on the device/component/system reliability, designs and applications in 45–90 nm era, because it may cause significant MBUs in memories which are not refreshed frequently or may cause MNTs in logic circuits that may corrupt space-redundancy soft-error mitigation techniques like triple module redundancy (TMR).

6.5 Countermeasures for Reliable Memory Design

This session describes a new architecture that corrects most multi-bit errors induced by cosmic ray. By the application of the approach in a 16-Mbit SRAM chip, fabricated in $0.13 - \mu m$ CMOS technology, the new ECC architecture reduces SERs by 99.5%.

6.5.1 ECC Error Correction and Interleave Technique for MCU

An on-chip error checking and correction (ECC) architecture is useful as a way of achieving a low SER. The conventional ECC circuit is, however, only capable of correcting a single error per address, and is thus incapable of handling cosmic ray-induced multi-errors.

To increase the rate of correction by the ECC circuit, a new architecture for the handling of cosmic ray-induced multi-errors are described. The key to this was to arrange the cells such that simultaneous cosmic ray-induced multiple errors are most likely to occur at different physical addresses. The concept and the effect of the developed physical address arrangement are shown.

A parasitic bipolar effect is responsible for such multiple errors [6.97], and each parasitic bipolar device operates in a single well. Multi-cell errors occur in cells which share the single well. In our lithographically symmetrical cell layout [6.85, 6.86], the p and n wells run parallel to the bit lines. The p-well is shared by the cells in adjacent columns. Therefore, multi-cell errors occur in the cells in adjacent columns. In other words, there are not more than two contiguous errors in the direction of word lines, while there are more than two contiguous errors in the direction of bit lines. This is consistent with the actual observations in the experimental results shown in the previous sections.

To increase the rate of correction by the ECC circuit in its current form, we developed a new architecture for the handling of cosmic ray-induced multi-errors. The concept of the physical address arrangement is shown in Fig. 6.27. The ECC circuit we designed uses a Hamming code to produce 128-bit data from 128 bits of data and ten parity bits. Each word line is formed of alternating cells, including data and parity cells, from two different addresses, e.g., address <0> and address <1> on Word <0>. From this address arrangement, we named this architecture the alternate ECC architecture. As a result, adjacent cells are for different addresses. All cells (128 data and ten parity) at an address are simultaneously read out to the ECC circuit and a single error at each address is correctable. The two cell errors induced by one neutron in "multi-error A" are corrected by the ECC circuit because the cells belong to different addresses. The three cell errors in "multi-error B" are corrected whether or not alternate ECC is used.

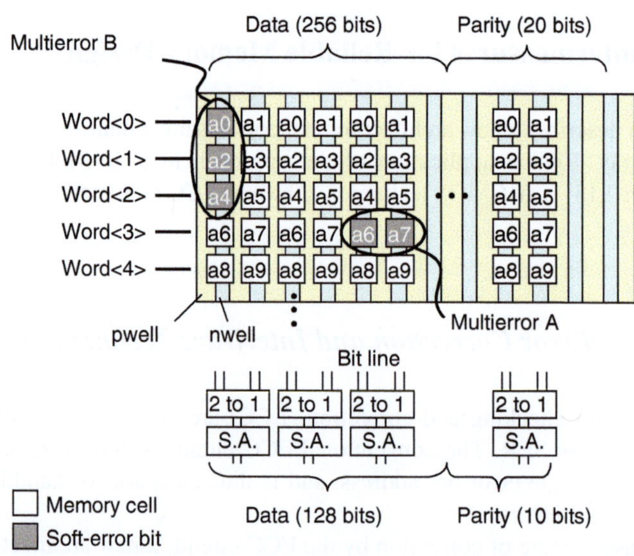

Fig. 6.27 Physical address arrangement for the handling of cosmic ray-induced multi-errors

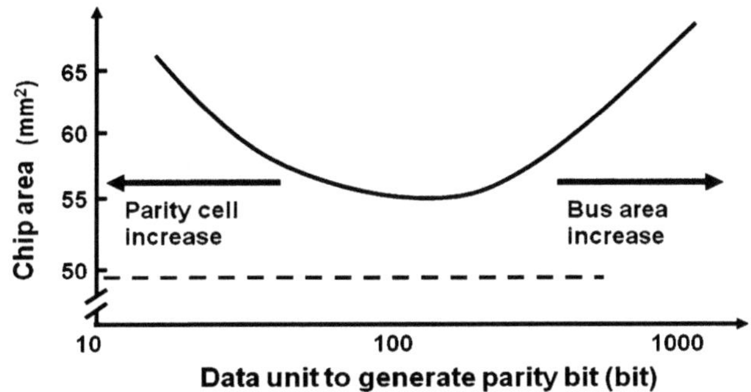

Fig. 6.28 Chip area as a function of data unit to generate parity bit

6.5.2 ECC Architecture

We investigate the architecture using ECC circuits. Access to 16 Mbit SRAM chip is in 16-bit units. Figure 6.28 shows the area overhead as a function of data width to generate parity bit using a Hamming code. Every 16-bit data needs the five parity bits. Many parity bits are needed, and the chip area overhead is very large. Although the 256-bit data needs only the nine parity bits, 265-bit buses are needed and area overhead increases. In this case, to minimize the area overhead, 128-bit data have the eight parity bits.

However, the write operation of 16-bit access causes the problem because 128-bit data are needed to generate parity bits. The new write operation is proposed as shown in Fig. 6.29. At write operation, at first, the 128-bit data and the eight parity bits are read out and the error bit is corrected. Next, 16-bit write data are replaced in the corrected 128-bit data. And the replaced 16-bit data regenerates the eight parity bits. Finally, the replaced 16-bit data and the eight parity bits are written in the cells. This method makes 16-bit access to chip possible even if the every 128-bit data has the eight parity bits.

Although access to the chip is in 16-bit units, 138-bit bus connections must span the whole chip to connect the sense-amplifier and ECC circuits. A hierarchical bus architecture was developed to reduce the power consumption and area overhead of the 138-bit bus. Two ECC circuits are placed on the chip. The overall data bus structure consists of four local 138-bit buses and two global 16-bit global buses, one for reading and the other for writing. Two of the local buses are connected to each ECC circuit, and both global buses are connected to both of the ECC circuits. The hierarchical architecture led to a 3% smaller chip and a 22% lower level of power dissipation. The total area penalty of the ECC circuits, parity cells, and 138-bit local buses is 9.7%. The chip's area is 56.2 mm², of which each ECC circuit only takes up 0.21 mm².

The effects of the techniques for reducing the area overhead are summarized in Fig. 6.30. The area increases by 36% if each 16-bit data have five parity bits

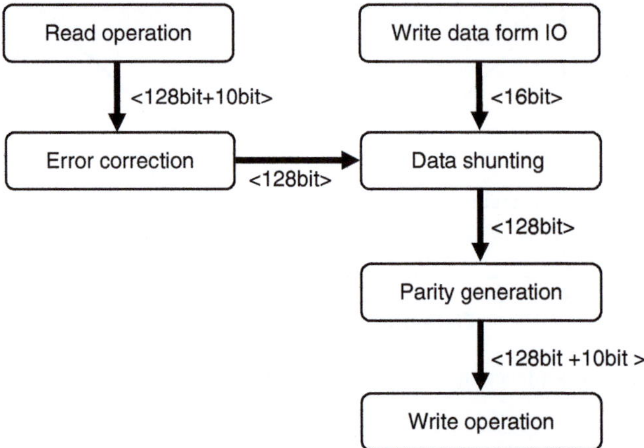

Fig. 6.29 Write method using ECC circuits

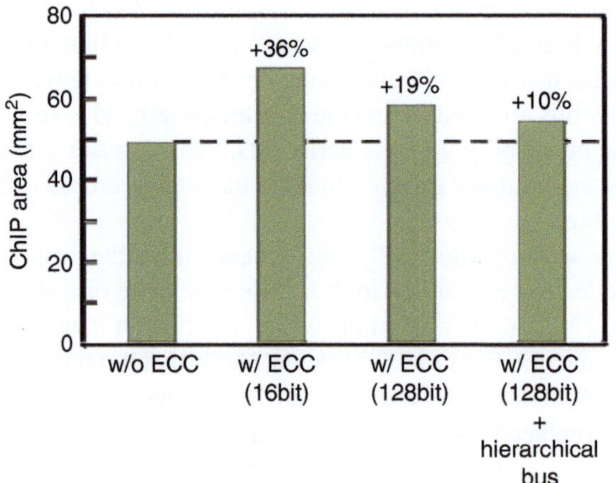

Fig. 6.30 Effect of proposed ECC circuits

compared to the area without parity bits. The area overhead reduces to 19% if each 128-bit data have eight parity bits. The hierarchical bus architecture reduces the area overhead to 10%.

6.5.3 Results

This 16 Mbit SRAM was fabricated by using three-metal 0.13-μm CMOS technology [6.98]. The process and device parameters are shown in Table 6.4. The gate length of the nMOS and pMOS devices is 0.14 μm. The threshold voltages of the

Table 6.4 Process and device parameters of 16Mb SRAM

Process	$3 - \text{metal } 0.13 - \mu\text{mCMOS}$
Threshold voltage	0.7 V: nMOS in memory cell
	1.0 V: pMOS in memory cell
	0.3 V: peripherals
Gate oxide	3.7 nm (electrical): internal
	8.4 nm (electrical): external
Cell size (6T)	$0.92 \times 2.24 \mu\text{m}$

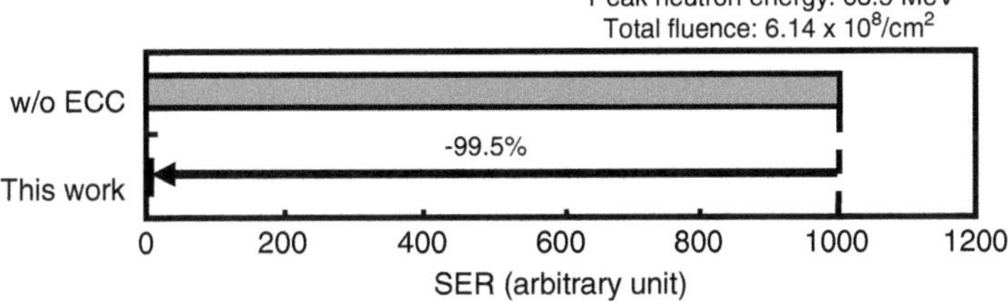

Fig. 6.31 Measured soft-error rates under neutron irradiation

memory cells are 0.7 V for nMOS, and 1 V for pMOS. On the other hand, the threshold voltage of the peripheral circuits is 0.3 V. The gate oxide layers of the internal circuits are 3.7 nm thick and those of the external circuits are 8.4 nm thick. The cell size is 0.92 by 2.24 μm. The gate width of the transfer nMOS and the load pMOS devices is 0.18 μm. The gate width of the driver nMOS device is 0.24 μm.

An SER acceleration test was carried out at the Cyclotron–Radio-Isotope Center (CYRIC). The peak neutron energy was 63.5 MeV and the total neutron fluence was $6.14 \times 10^8/\text{cm}^2$. The measured results for SER are shown in Fig. 6.31. The induction of multi-cell errors at the same address by a single neutron incidence was not seen, even once, in the 2,000 error events. As a result, the alternate ECC architecture had reduced the SER by 99.5% from the result for the SRAM chips with no ECC. The decrease ratio in SER of the alternate ECC scheme to a conventional ECC scheme depends on the data pattern. The maximum decrease ratio almost corresponds to 99.5%.

References

6.1. E.L. Petersen, The SEU figure of merit and proton upset rate calculations IEEE Trans. Nucl. Sci. **45**(6), 2550–2562 (1998)

6.2. P.E. Dodd et al., Impact of ion energy on single-event upset. IEEE Trans. Nucl. Sci **45**(6), 2483–2491 (1998)

6.3. T. Goka, H. Matsumoto, N. Nemoto, SEE flight data from Japanese satellites. IEEE Trans. Nucl. Sci. **45**(6), 2771–2778 (1998)

6.4. R. Koga, Single event sensitivities of 128- and 256-megabit synchronous dynamic random access memories (SDRAMs), in *Proceedings of the 4th International Workshop on Radiation Effects on Semiconductor Devices for Space Application*, Tsukuba, 15–18 May 2000, pp. 81–88

6.5. E. Normand Extensions of burst generation rate method for wider application to proton/neutron-induced single event effects. IEEE Trans. Nucl. Sci **45**(6), 2904–2914 (1998)

6.6. E. Normand, T.J. Baker, Altitude and latitude variations in avionics SEU and atmospheric neutron flux IEEE Trans. Nucl. Sci. **40**(6), 1484–1490 (1993)

6.7. T. Nakamura, Y. Uwamino, T. Ohkubo, A. Hara, Altitude variation of cosmic-ray neutrons. Health Phys. **53**(5), 509–517 (1987)

6.8. T.C. May, M.H. Woods, Alpha-particle-induced soft errors in dynamic memories. IEEE Trans. Electron Devices **ED-26**(1), 2–9 (1979)

6.9. J.F. Ziegler, W.A. Lanford, Effect of cosmic rays on computer memories. Science **206**(4420), 776–788 (1979)

6.10. E. Takeda, K. Takeuchi, E. Yamazaki, T. Toyabe, K. Ohshima, K. Itoh, Effective funneling length in alpha-particle induced soft errors in *Extended Abstracts of the 18th Conference on Solid State Devices and Materials*, Tokyo, pp. 311–314 (1986)

6.11. A. Eto, M. Hidaka, Y. Okuyama, K. Kimura, M. Hosono, impact of neutron flux on soft errors in MOS memories, *International Electron Devices Meeting*, San Francisco, CA, December 6–9, pp. 367–370 (1998)

6.12. D.C. Bossen, CMOS soft errors and server design, in *Workshop on Radiation Induced Soft Errors in Silicon Components and Computer Systems, IRPS*, Dallas, 7 April 2002

6.13. C. Slayman, Eliminating the threat of soft errors – a system vendor perspective, in *IRPS Panel Discussion on Eliminating the Threat of Soft Errors*, Dallas, 2 April 2003, No. 6

6.14. E. Normand, Single event upset at ground level. IEEE Trans. Nucl. Sci. **43**(6), 2742–2750 (1996)

6.15. E. Ibe, Current and future trend on cosmic-ray-neutron induced single event upset at the ground down to 0.1-micron-device, in *TSL Workshop on Applied Physics*, Uppsala, 3 May 2001, No. 1

6.16. E. Ibe, Y. Yahagi, F. Kataoka, Y. Saito, A. Eto, M. Sato, H. Kameyama, M. Hidaka, A self-consistent integrated system for terrestrial-neutron induced single event upset of semiconductor devices at the ground, in *2002 ICITA*, Buthurst, No. 273–21 (2002)

6.17. T. Nakamura, M. Baba, E. Ibe, Y. Yahagi, H. Kameyama, *Terrestrial Neutron-Induced Soft-Errors in Advanced Memory Devices* (World Scientific, New Jersey, 2008)

6.18. L. Borucki, G. Schindlbeck, C. Slayman, Comparison of accelerated DRAM soft error rates measured at component and system level, in *IRPS 2008*, Phoenix Convention Center, Phoenix, 27 April – 1 May 2008 No. 5A.4

6.19. G. Gasiot, D. Giot, P. Roche, alpha-induced multiple cell upsets in standard and radiation hardened SRAMs manufactured in a 65 nm CMOS technology (TNS). Trans. Nucl. Sci **53**(6), 3479–3486 (2006)

6.20. P. Shivakumar, M. Kistler, S.W. Keckler, D. Burger, A. Lorenzo, Modeling the effect of technology trends on the soft error rate of combinational logic, in *International Conference on Dependable Systems and Networks*, pp. 389–398 (2002)

6.21. N. Seifert, X. Zhu, L. Massengill, Impact of scaling on soft-error rates in commercial microprocessors. Trans. Nucl. Sci. **49**(6), 3100–3106 (2002)

6.22. S. Mitra, M. Zhang, S. Waqas, N. Seifert, B. Gill, K.S. Kim, Combinational logic soft error correction, in *The Second Workshop on System Effects of Logic Soft Errors*, Urbana-Champain, 11–12 April 2006

6.23. S. Hareland et al., Impact of CMOS process scaling and SOI on the soft error rates of logic processes, in *Symposium on VLSI Technology Digest of Technical Papers*, pp. 73–74 (2001)

6.24. M. Baze, J. Wert, J. Clement, M. Hubert, A. Witulski, Propagating SET characterization technique for digital CMOS libraries, in *International Nuclear and Space Radiation Effects Conference*, Ponte Vedra Beach, 17–21 July 2006, No. E4

6.25. N. Seifert, B. Gill, M. Zhang, V. Zia, V. Ambrose, On the scalability of redundancy based SER mitigation schemes, in *ICICDT2007*, Austin, 18–20 May 2007, No. G2, pp. 197–205

6.26. P.E. Dodd, M.R. Shaneyfelt, D.S. Walsh, J.R. Schwank, G.L. Hash, R.A. Loemker, B.L. Draper P.S. Winokur, Single-event upset and snapback in silicon-on-insulator devices and integrated circuits. Trans. Nucl. Sci. **47**(6), 2165–2174 (2000)

6.27. Schwank, J.R., Dodd, P.E., Felix, J.A., Sexton, F.W., Hash, G.L., M.R. Shaneyfelt, J. Baggio, V. Ferlet-Cavrois, P. Paillet, E. Blackmore, Effects of particle energy on proton and neutron-induced single-event latchup, in *2005 IEEE Nuclear and Space Radiation Effects Conference*, Seattle, 11–15 July 2005, No. I-4

6.28. P.E. Dodd, M.R. Shaneyfelt, J.R. Schwank, G.L. Hash, Neutron-induced softerrors, latchup, and comparison of SER test methods for SRAM technologies, in *IEDM*, No. 13.2 (2002)

6.29. R. Koga, S. Penzin, K. Crawford, W. Crain, Single event functional interrupt (SEFI) sensitivity in microcircuits in *Proceedings of RADECS 1997*, pp. 311–318 (1997)

6.30. Actel, Understanding soft and firm errors in semiconductor devices (2003), www.actel.com

6.31. S Huang, G.A.J. Amaratunga, Analysis of SEB and SEGR in super-junction MOSFETs. Trans. Nucl. Sci. **47**(6) 2640–2647 (2000)

6.32. S. Kuboyama, K. Sugimoto, S. Shugyo, S. Matsuda, T. Hirao, Single-event burnout of epitaxial bipolar transistors. Trans. Nucl. Sci **45**(6), 2527–2533 (1998)

6.33. E. Normand, J.L. Wert, D.L. Oberg, P.P. Majewski, P. Voss, SA Wender, Neutron-induced single event burnout in high voltage electronics. Trans. Nucl. Sci **44**, 2358–2368 (1997)

6.34. Semiconductor Industry Association, Technology Node Challenges, *International Technology Roadmap for Semiconductors*, 1999 Edition (1999)

6.35. Y. Yahagi, E. Ibe, S. Yamamoto, Y. Yoshino, M. Sato, Y. Takahashi, H. Kameyama, A. Saito, M. Hidaka, Versatility of SEU function and its derivation from the irradiation tests with well-defined white neutron beams. Trans. Nucl. Sci **52**(5), 1562–1567 (2005)

6.36. Y. Yahagi, E. Ibe, Y. Saito, A. Eto, M. Sato, *"Self-Consistent Integrated System for Susceptibility to Terrestrial-Neutron Induced Soft-error of Sub-quarter Micron Memory Devices, 2002 IRW* (Stanford Sierra Camp, South Lake Tahoe, 2002), pp. 143–146

6.37. Y. Yahagi, E. Ibe, Y. Takahashi, Y. Saito, A. Eto, M. Sato, H. Kameyama, M. Hidaka, K. Terunuma, T. Nunomiya, T. Nakamura, Threshold energy of neutron-induced single event upset as a critical factor, in *2004 IEEE International Reliability Physics Symposium*, Phoenix, 25–29 April 2004, pp. 669–670

6.38. JESD89A, Measurement and reporting of alpha particles and terrestrial cosmic ray-induced soft errors in semiconductor devices, Revision of JEDEC Standard No. 89, October 2006

6.39. J. Maiz, S. Hareland, K. Zhang, P. Armstrong, Characterization of multi-bit soft error events in advanced SRAMs, in *Technical Digest of IEEE International Electron Device Meeting (IEDM) 2003*, Washington, 7–10 December 2003, No. 21.4

6.40. D. Giot, G. Gasiot, P. Roche, Multiple bit upset analysis in 90 nm SRAMs, in *9th European Workshop on Radiation Effects on Components and Systems*, Athens, 27–29 September 2006, No.A-7, pp. 26–29

6.41. R. Baumann, The impact of technology scaling on soft error rate performance and limits to the efficacy of error correction, in *Inter. Electron Devices Meeting (IEDM) Tech. Digest*, San Francisco, 8–11 December 2002, pp. 329–332

6.42. T.C. May, M.H. Woods, A new physical mechanism for soft errors in dynamic memories, in *16th Annual Proceedings of 1978 International Reliability Symposium*, 18–20 April, San Diego, 1979, pp. 33–40

6.43. T.C. May, M.H. Woods, Alpha-particle-induced soft errors in dynamic memories. IEEE Trans. Electron Devices **ED-26**(1), 2–9 (1979)

6.44. D.S. Yaney, J.T. Nelson, L.L. Vanskike, Alpha-particle tracks in silicon and their effect on dynamic MOS RAM reliability. IEEE Trans. Electron Devices **ED-26**(1), 10–16 (1979)

6.45. M. Koyanagi, H. Sunami, N. Hashimoto, M. Ishikawa, Novel high density, stacked capacitor MOS RAM, in *Inter. Electron Devices Meeting (IEDM) Tech. Digest*, Washington, 4–6 December 1978, pp. 348–351

6.46. H. Sunami, T. Kume, N. Hashimoto, K. Ito, T. Toyabe, S. Asai, A corrugated capacitor cell (CCC) for megabit dynamic MOS memories, in *Inter. Electron Devices Meeting (IEDM) Tech. Digest*, San Francisco, 13–15 December 1982, pp. 806–809

6.47. H. Shinnriki, T. Kisu, S. Kimura, Y. Nishioka, Y. Kawamoto, K. Mukai, Promising storage capacitor structures with thin Ta2O5 film low-power high-density DRAM's. IEEE Trans. Electron Devices **37**(9), 1939–1947 (1990)

6.48. Y. Nishioka, H. Shinriki, K. Mukai, Influence of SiO2 at the Ta2O5/Si interface on dielectric characteristics of Ta2O5 capacitors. J. Appl. Phys. **61**(6), 2335–2338 (1987)

6.49. S. Takehiro, S. Yamauchi, M. Yoshimaru, H. Onoda, The simple stacked BST capacitor for the future DRAMs using a novel low temperature growth enhanced crystallization, in *Digest of 1997 Symposium on VLSI Technology*, Kyoto, June 10–12, pp. 153–154 (1997)

6.50. Y. Kohyama, T. Ozaki, S. Yoshida, Y. Ishibashi, H. Nitta, S. Inoue, K. Nakamura, T. Aoyama, K. Imai, H. Hayasaka, A fully printable, self-aligned and planarized stacked capacitor DRAM cell technology for 1 Gbit DRAM and beyond, in Digest. of 1997 Symposium on VLSI Technology, Kyoto, June 10–12, pp. 17–18, (1997)

6.51. Y. Tosaka, K. Suzuki, S. Satoh, T. Sugii, Theoretical study of alpha-particle-induced soft errors in submicron SOI SRAM. IEICE Trans. Electron. **E79-C**(6), 767–771 (1996)

6.52. Y. Tosaka, S. Satoh, T. Itakura, K. Suzuki, T. Sugii, H. Ehara, G.A. Woffinden, Cosmic ray neutron-induced soft errors in sub-half micron CMOS circuits. IEEE Electron Device Lett. **18**(3), 99–101 (1997)

6.53. Y. Tosaka, S. Satoh, T. Itakura, H. Ehara, T. Ueda, G.A. Woffinden, S.A. Wender, Measurement and analysis of neutron-induced soft errors in sub-half micron CMOS circuits. IEEE Trans. Electron Devices **45**(7), 1453–1458 (1998)

6.54. Y. Hirano, T. Iwamura, K. Shiga, K. Nii, K. Sonoda, T. Matsumoto, S. Maeda, Y. Yamaguchi, T. Ipposhi, S. Maegawa, Y. Inoue, High soft-error tolerance body-tied SOI technology with partial trench isolation (PTI) for next generation devices, in *2002 Symposium on VLSI Technology: Digest of Technical Papers*, 11–13 June, Honolulu, IEEE CAT. No. 01CH37303, JSAP CAT. No. AP021201, pp. 48–49, (2002)

6.55. H. Masuda, T. Toyabe, H. Shukuri, K. Ohshima, K. Itoh, A full three-dimensional simulation on alpha-particle induced DRAM soft errors, in *Technical Digest – International Electron Devices Meeting*, pp. 496–499 (1985)

6.56. C.M. Hsieh, P.C. Murley, R.R. O'Brien, A field-funneling effect on the collection of alpha-particle-generated carriers in silicon devices. IEEE Electron Device Lett. **EDL-2**(4), 103–105 (1981)

6.57. P.M. Carter, B.R. Wilkins, Influences on soft error rates in static RAM's. IEEE J. Solid-State Circuits **SC-22**(3), 430–436 (1987)

6.58. S. Satoh, R. Sudo, H. Tashiro, N. Higaki, N. Nakayama, CMOS-SRAM soft-error simulation system, in *1994 IEEE/IPRS*, p. 339

6.59. K. Takeuchi, K. Shimohigashi, E. Takeda, E. Yamasaki, T. Toyabe, K. Itoh, Alpha-particle-induced charge collection measurements for megabit DRAM cells. IEEE Trans. Electron Devices **ED-36**(9), 1644–1650 (1989)

6.60. K. Takeuchi, K. Shimohigashi, H. Kozuka, T. Toyabe, K. Itoh, H. Kurosawa, Origin and characteristics of alpha-particle-induced permanent junction leakage. IEEE Trans. Electron Devices **ED-37**(3), 730–736 (1990)

6.61. E. Takeda, K. Takeuchi, E. Yamasaki, T. Toyabe, K. Ohshima, K. Itoh, The scaling law of alpha-particle induced soft errors for VLSI's, in *IEDM '86*, pp. 542–545 (1986)

6.62. H. Shin, Modeling of alpha-particle-induced soft error rate in DRAM. IEEE Trans. Electron Devices **46**(9), 1850–1857 (1999)

6.63. M. Minami, Y. Wakui, H. Matsuki, T. Nagano, A new soft-error-immune static memory cell having a vertical diver MOSFET with a buried source for ground potential. IEEE Trans. Electron Devices **36**(9), 1657–1662 (1989)

6.64. K. Yamaguchi, Y. Takemura, K. Osada, K. Ishibashi, Y. Saito, Three-dimensional device modeling for SRAM soft-error immunity and tolerance analysis. IEEE Trans. Electron Devices **51**(3), 378–388 (2004)

6.65. D. Leroy, R. Gaillard, E. Schaefer, C. Beltrando, S.-J. Wen, R. Wong, Variation of alpha induced soft error rate with technology node, in *IOLTS 2008*, Greece, 6–9 July 2008, No.11.2, pp. 253–260

6.66. E.P. Rech, S. Gerardin, A. Paccagnella, P. Bernardi, M. Grosso, M. Sonza Reorda, D. Appello, Evaluating alpha-induced soft errors in embedded microprocessors, in *IOLTS2009*, Sesimbra-Lisbon, 24–26 June 2009, No. 4.1, pp. 69–74

6.67. E. Ibe, Y. Yahagi, H. Kameyama, Y. Takahashi, Single event effects of semiconductor devices at the ground. Ionizing Radiation **30**(7), 263–281, (2004)

6.68. M.S. Gordon, P. Goldhagen, K.P. Rodbell, T.H. Zabel, H.H.K. Tang, J.M. Clem, P. Bailey, Measurement of the flux and energy spectrum of cosmic-ray induced neutrons on the ground. IEEE Trans. Nucl. Sci. **51**, 3427–3434 (2004)

6.69. C. Hu, Alpha-particle-induced field and enhanced collection of carriers. IEEE Electron Device Lett **EDL-3**(2), 31–34 (1982)

6.70. M. Baba, M. Takada, T. Iwasaki, S. Matsuyama, T. Nakamura et al, Development of monoenergetic neutron calibration fields between 8 keV and 15 MeV. Nucl. Instrum. Methods Phys. Res. A **376**, 115–123 (1996)

6.71. M. Baba, H. Okamura, M. Hagiwara, T. Itoga, S. Kamada, Y. Yahagi, E. Ibe, Installation and application of an intense ^{7}Li(p, n) neutron source for 20–90 MeV region. Radiat. Prot. Dosimetry **126**(1–4), 13–17 (2007)

6.72. M. Baba, Installation and application of an intense ^{7}Li(p, n) neutron source for 20–90 MeV region, in *NEUDOS-10*, Uppsala, 12–16 June 2006, No. A1–3

6.73. A.V. Prokofiev, O. Bystrom, C. Ekstrom, D. Reistad, D. Wessman, S. Pomp, J. Blomgren, M. Osterlund, U. Tippawan, V. Ziemann, A new neutron beam facility for SEE testing, in *2005 RADECS, Sep.19–23, Palais des Congres*, Cap d'Agde, 2005, No.W-14

6.74. E. Ibe, Y. Yahagi, H. Yamaguchi, H. Kameyama, SEALER: novel Monte-Carlo simulator for single event effects of composite-materials semiconductor devices, in *2005 RADECS, Sep. 19–23, Palais des Congres*, Cap d'Agde, 2005, No. E-4

6.75. T. Nakamura, T. Ninomiya, N. Hirabayashi, H. Suzuki, Y. Sato, Sequential measurements of energy spectrum and intensity for cosmic ray neutrons, in *Proceedings of the 7th International Symposium on Natural Radiation Environment (NRE-VII)*, Rhodes, 20–24 May 2002

6.76. H.W. Bertini, A.H. Culkowski, O.W. Hermann, N.B. Gove, M.P. Guthrie, High energy($E \leqq$ 100 GeV) intranuclear cascade model for nucleons and pions incident on nuclei and comparisons with experimental data. Phys. Rev. C **17**(4) 1382–1394 (1978)

6.77. I. Dostrovsky, Z. Fraenkel, G. Friedlander, Monte Carlo calculations of nuclear evaporation process. III. Applications to low-energy reactions. Phys. Rev. **116**(3), 683–702 (1959)

6.78. F. Bertland, R. Peele, Complete hydrogen and helium particle spectra from 30- to 60-MeV proton bombardment of nuclei with $A = 12 \, to \, 209$ and comparison with the intranuclear cascade model. Phys. Rev. C Nucl. Phys. **8**(3), 1045–1064 (1973)

6.79. H.H.K. Tang, G.R. Srinivasan, N. Azziz, Cascade statistical model for nucleon-induced reactions on light nuclei in the energy range 50-MeV-1 GeV. Phys. Rev. C **42**(4), 1598–1622 (1990)

6.80. S. Satoh, Y. Tosaka, T. Itakura, Scaling law for secondary cosmic-ray neutron-induced soft errors in DRAMs, in *Extended Abstracts of the 1998 ISDM, Hiroshima*, pp. 40–41 (1998)

6.81. R.C. Baumann, E.B. Smith, Neutron-induced boron fission as a major source of soft errors in deep submicron SRAM devices, in *2000 IRPS*, San Jose, 10–13 April 2000, pp. 152–157

6.82. E. Ibe, H. Taniguchi, Y. Yahagi, K. Shimbo, T. Toba, Impact of scaling on neutron-induced soft error in SRAMs from a 250 nm to a 22 nm design rule. Trans. Electron Device **57**(7), 1527–1538, (2010)

6.83. J. Yamada, Selector-line merged built-in ECC technique for DRAM's. IEEE J. Solid-State Circuits **22**(5), 868–873 (1987)

6.84. K. Furutani, K. Arimoto, H. Miyamoto, T. Kobayashi, K. Yasuda, K. Mashiko, A built-in hamming code ECC circuit for DRAM's. IEEE J. Solid-State Circuits **24**(1), 50–56 (1989)

6.85. E. Ibe, S. Chung, S. Wen, Y. Yahagi, H. Kameyama, S. Yamamoto, T. Akioka, H. Yamaguchi, Valid and prompt track-down algorithms for multiple error mechanisms in neutron-induced single event effects of memory devices, in *RADECS*, Athens, 27–29 September 2006, No. D-2

6.86. E. Ibe, H. Kameyama, Y. Yahagi, K. Nishimoto, Y. Takahashi, Distinctive asymmetry in neutron-induced multiple error patterns of 0.13 μm process SRAM, in *The 6th International Workshop on Radiation Effects on Semiconductor Devices for Space Application*, Tsukuba, 6–8 October 2004, pp. 19–23

6.87. E. Ibe, S. Chung, S. Wen, H. Yamaguchi, Y. Yahagi, H. Kameyama, S. Yamamoto, T. Akioka, Spreading diversity in multi-cell neutron-induced upsets with device scaling, in *2006 CICC*, San Jose, 10–13 September 2006, pp. 437–444

6.88. E. Ibe, S.S. Chung, S.J. Wen, H. Yamaguchi, Y. Yahagi, H. Kameyama, S. Yamamoto, Multi-error propagation mechanisms clarified in CMOSFET SRAM devices under quasi-mono energetic neutron irradiation, in *NSREC2006*, Ponte Vedra Beach, No. PC-6 (2006)

6.89. K. Osada, J.L. Shin, M. Khan, Y. Liou, K. Wang, K. Shoji, K. Kuroda, S. Ikeda, K. Ishibashi, Universal-Vdd 0.65–2.0 V 32 kB cache using voltage-adapted timing-generation scheme and a lithographical-symmetric cell, in *Proceedings of the IEEE International Solid-State Circuits Conference*, February 2001, pp. 168–169

6.90. K. Osada, J.L. Shin, M. Khan, Y. Liou, K. Wang, K. Shoji, K. Kuroda, S. Ikeda, K. Ishibashi, Universal-V_{dd} 0.65–2.0-V 32-kB cache using voltage-adapted timing-generation scheme and a lithographical-symmetric cell. IEEE J. Solid-State Circuits **36**(11), 1738–1744 (2001)

6.91. T. Toyabe, H. Masuda, Y. Aoki, H. Shukuri, T. Hagiwara, Three-dimensional device simulator CADDETH with highly convergent matrix solution algorithms. IEEE Trans. Electron Device **32**(10), 2038–2044 (1985)

6.92. S.M. Sze, KK. Ng, *Physics of Semiconductor Devices*, 3rd edn. (Wiley Interscience, San Jose, 2006), pp. 7–133

6.93. S.M. Sze, KK. Ng, *Physics of Semiconductor Devices*, 3rd edn. (Wiley Interscience, San Jose, 2006), pp. 249–250

6.94. C. Lombardi, S. Manzini, A. Saporito, M. Vanzi, A physically based mobility model for numerical simulation of nonplanar devices. IEEE Trans. CAD **7**, 1164–1171 (1988)

6.95. M. Valdinoci, D. Ventura, M.C. Vecchi, M. Rudan, G. Baccarani, F. Illien, A. Stricker, L. Zullino, Impact-ionization in silicon at large operating temperature, in *International Conference on Simulation of Semiconductor Processes and Devices*, September 1999

6.96. H. Yamaguchi, E. Ibe, Y. Yahagi, S. Yamamoto, T. Akioka, H. Kameyama, Novel mechanism of neutron-induced multi-cell error in CMOS devices tracked down from 3D device simulation, in *Proceedings of the 2006 International Conference on Simulaton of Semiconductor Process and Devices (SISPAD)*, Monterey, 6–8 September 2006, pp. 184–187

6.97. K. Osada, K Yamaguchi, Y. Saitoh, T. Kawahara, Cosmic-ray multi-error immunity for SRAM, based on analysis of the parasitic bipolar effect, in *Symposium on VLSI Circuits Digest*, June 2003, pp. 255–256

6.98. K. Osada, Y. Saitoh, E. Ibe, K. Ishibashi, 16.7-fA/cell tunnel-leakage-suppressed 16-Mbit SRAM for handling cosmic-ray-induced multi-errors IEEE J. Solid-State Circuits **38**(11), 1952–1957 (2003)

Chapter 7
Future Technologies

Koji Nii and Masanao Yamaoka

The design solution described in Chap. 5 will help the minimum operating voltage (VDD_{min}) ofa general 6T single-port SRAM. However, it will eventually face the limitation of the SRAM VDD_{min} because of the degradation of the SRAM stability due to an increase in the local V_{th} variation. In this chapter, first some alternative 6T single-port SRAM cells to enhance the SRAM stability are introduced in Sect. 7.1.

By the end of Moore's low, 3D channel transistors are researched and developed, and will be the main stream. The SRAM cells using 3D transistors are described. In Sect. 7.2, a low V_{th} variation transistor, thin-BOX FD-SOI transistor, is proposed. The six- and four-transistor memory cells composed of the thin-BOX FD-SOI transistor can achieve the low-cost and high-performance SRAM modules. In Sect. 7.3, SRAM cell using FinFET is described. The FinFET has 3D channel and break through the short channel problem.

7.1 7T, 8T, 10T SRAM Cell

Alternative 6T single-port SRAM cell designs, which use 7T or 8T read margin-free cell or several types of 10T read margin-free cells, are reported recently [7.1, 7.2, 7.3]. Figure 7.1 depicts the typical types of schematics for an alternative 6T SRAM cell.

To improve the read margin of an SRAM, which is the static noise margin (SNM), additional transistors areused so as not to disturb the storage node during the read operation. A 7T SRAM cell was proposed to improve the read margin

K. Nii (✉)
Renesas Electronics Corporation, 5-20-1, Josuihon-cho, Kodaira, Tokyo 187-8588, Japan
e-mail: koji.nii.uj@renesas.com

M. Yamaoka
1101 Kitchawan Road, Yorktown Heights, NY 10598, USA
e-mail: masanao.yamaoka@hal.hitachi.com

K. Ishibashi and K. Osada (eds.), *Low Power and Reliable SRAM Memory Cell and Array Design*, Springer Series in Advanced Microelectronics 31, DOI 10.1007/978-3-642-19568-6_7, © Springer-Verlag Berlin Heidelberg 2011

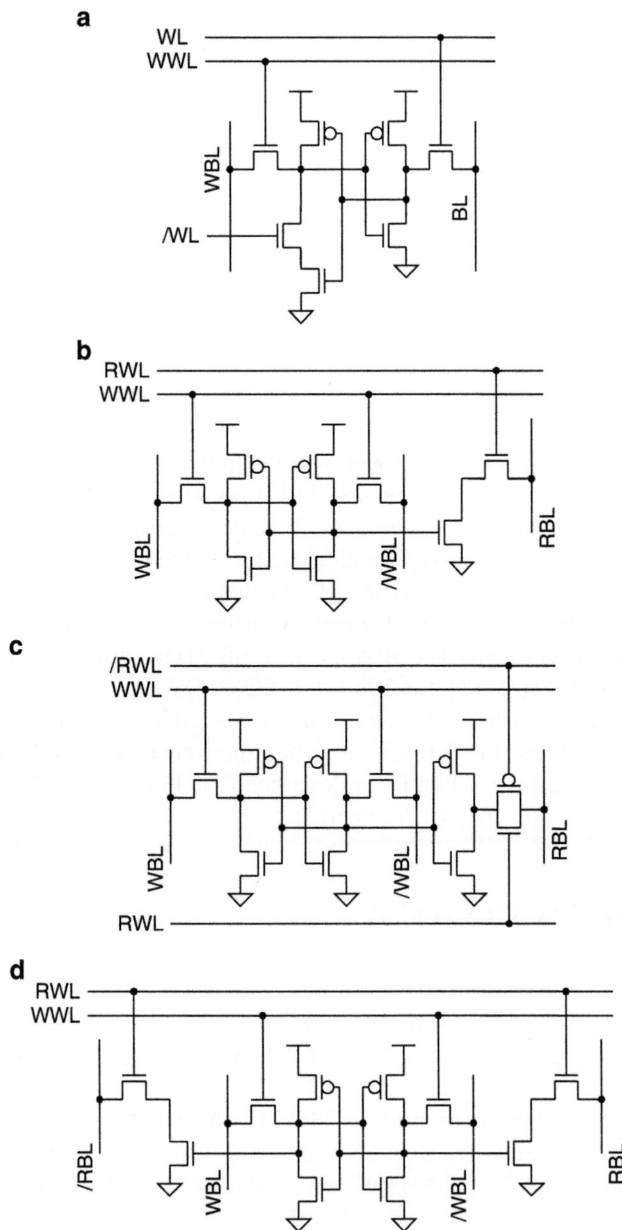

Fig. 7.1 (continued)

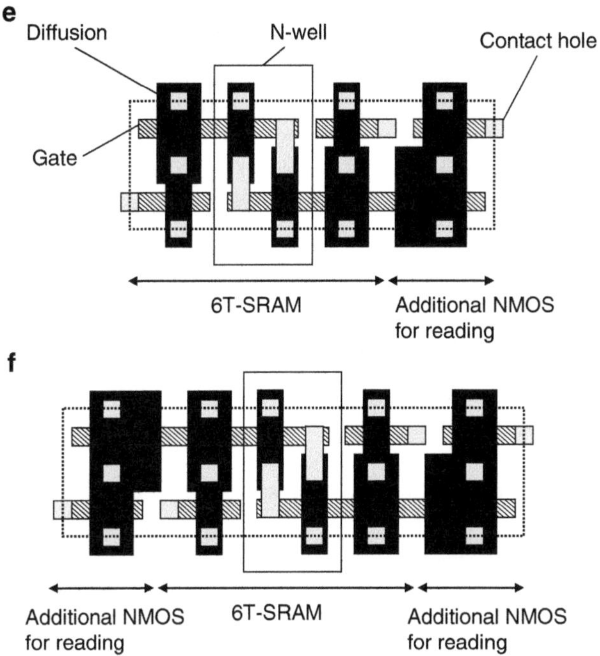

Fig. 7.1 Alternative 6T SRAM bitcells, (**a**) 7T, (**b**) 8T single-end WBL, (**c**) 10T single-end, (**d**) 10T double-end, (**e**), (**f**) layouts of 8T single-end and 10T double-end

by an additional pull-down NMOS [7.1] as shown in Fig. 7.1a. Typically, an 8T SRAM cell, whose schematic is depicted in Fig. 7.1b, becomes widely used for advanced CMOS devices. This type of memory cell has two additional NMOS transistors with a single read bitline (RBL) for read operation. For write operations, complementary write bitline pairs (WBL, /WBL) are used just as in the 6T SRAM cell. Consequently, this 8T SRAM cell improves read stability over that of the 6T SRAM cell. The 8T SRAM cell area would eventually increase because of the two additional NMOS transistors if the same gate length and width were used for each memory cell transistor. By considering the local V_{th} variation, the transistor sizes of the 6T SRAM cannot shrink further. The memory transistor sizes must be designed to become bigger than those of the previous generation to ensure the cell stability in sub 50 nm technology. Meanwhile, the 8T SRAM cell has strong robustness of read stability against process variation. For that reason, it can shrink the transistor size in accordance to the scaling trend for 45, 32 or 22 nm generations. This might induce 6T SRAM cell size becoming larger than the 8T SRAM cell. The crossover point of this area advantage for future generations is predicted and discussed in some reports [7.4, 7.5].

On the other hand, a non-precharge type of 10T SRAM cell is proposed as depicted in Fig. 7.1c. This type of memory cell has a read buffer within a cell, just as the 8T SRAM cell does. The pull-up and transfer PMOS transistors and a complementary read wordline pair (RWL, /RWL) is used to achieve the precharge less RBL structure. Although this 10T SRAM cell has a larger area penalty than that of 8T SRAM cell, the active power can be reduced further depending on the toggle rate of readout data. Thereby, it becomes more effective in power reduction under the low-voltage operation. However, these single-ended bitline structures, which operate at a full voltage swing for RBL, present the disadvantage that the transition time of RBL has large variation because of local V_{th} variation. To make matters worse, the former 8T SRAM cell requires a keeper circuit for RBL, conflicting the readout data. The 10T SRAM ladder cell also presents the disadvantage that the rising time of RBL slows because of the weak drive strength of the serially connected pull-up PMOSs.

The SNM-free 10T SRAM cell with four extra NMOS transistors, which is depicted in Fig. 7.1d, has a differential read bitline pair (RBL, /RBL). This memory cell can compensate the variation of readout transition time because the small differential signal can be detected using a sense amplifier as well as sensing of 6T SRAM cell. The area penalty is almost the same as that of the 10T SRAM cell, as depicted in Fig. 7.1c.

The layout plots of 8T SRAM cell and 10T SRAM cell are depicted in Fig. 7.1e, f. These layouts can be implemented by adding the additional NMOS to the typical 6T SRAM bitcell. The gates for PU, PD, PG, and additional NMOS are arranged in all same direction. Typically, the area of 8T SRAM cell becomes 1.3 times larger than the 6T SRAM cell, and that of 10T SRAM cell becomes 1.6 times.

These typical alternative 6T SRAM cells present advantages of the read stability. Notwithstanding, the half select issue during writing operation remains. This half select issue occurs in cells with a selected row and unselected column in the case of column multiplexing.

7.2 Thin-Box FD-SOI SRAM

The major problem of bulk MOSFETs in process scaling is increasing V_{th} variation. A low V_{th} variation transistor, a thin-BOX FD-SOI (thin buried oxide, fully depleted silicon on insulator) MOSFET, is proposed. The channel dopant of this MOSFET is very thin, and the V_{th} variation is controlled to low level. This thin-BOX FD-SOI MOSFET also has other prominent characteristics; a back-gate controllability without extra leakage current. Using this back-gate controllability, several circuit techniques have been proposed, and achieve the low-cost and high-performance SRAM modules. This collaboration between circuit level and device level technique has a potential to break through the transistor's scaling limit by V_{th} variation.

There are two options to operate the small-size memory cells at a low V_{dd}. One is to improve the operating margin and the other is to suppress the increasing V_{th} variations itself. Many circuit techniques to improve operating margins are proposed. In this chapter, the other solution to these problems is proposed. It starts with a new transistor structure, thin-BOX FD-SOI (fully depleted SOI). Figure 7.2 shows two cross-sectional views of this proposed transistor, the left in the direction

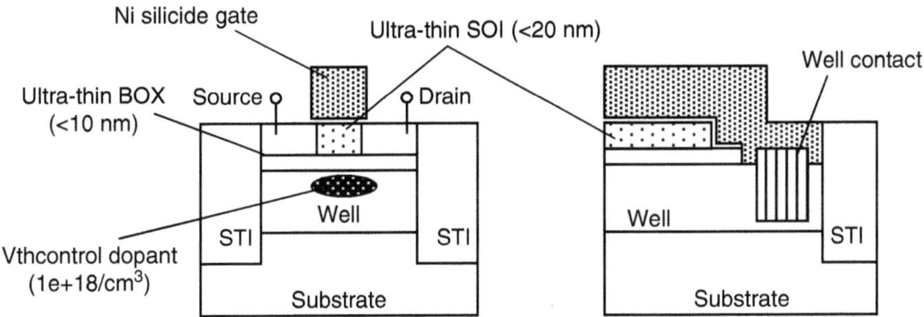

Fig. 7.2 Cross-sectional views of thin-BOX FD-SOI transistor

of gate width and the right in the direction of gate length. This thin-BOX FD-SOI transistor has several special features: an ultrathin SOI layer (<20 nm) and an ultrathin-BOX (buried oxide) layer (<10 nm). The FD-SOI structure controls the local component of V_{th} variations V_{th} variations by almost completely eliminating fluctuation of channel dopant distributions. An Ni silicide gate (mid-gap work-function metallic gate) and a deep impurity dopant under channel are used to provide a suitable V_{th} value for usage in low-power circuits. The substrate under the BOX layer is doped by means of implantation through the ultrathin SOI and BOX layers. The V_{th} is then mainly determined by work function of gate electrode; this reduces the effects of fluctuations in L_g, W, T_{ox}, etc. and thus controls the global components of V_{th} variations.

Figure 7.3 shows the V_{th} variation of conventional bulk transistors and the proposed thin-BOX FD-SOI transistors. The upper four bars indicate the standard deviation of the V_{th} variation for bulk transistors, and the lower three bars indicate the predicted standard deviation of the V_{th} variations for thin-BOX FD-SOI transistors. The V_{th} variation of bulk transistors is growing with the process scaling enormously; on the other hand, the growth of the V_{th} variation of FD-SOI is small, and the standard deviation of 32 nm V_{th} variation of FD-SOI is still smaller than that of 90 nm bulk transistor. Applying low V_{th} variation transistor to SRAM memory cell, the operating window becomes wide. Figure 7.4 shows 0.7 V V_{th} window curves of SRAM cells with bulk and FD-SOI transistors. It shows that the memory cell with FD-SOI can operate with a supply voltage of below 0.7 V, even when IR-drops and power-source fluctuations are taken into account.

Figure 7.5 is a cross-sectional SEM micrograph of a fabricated thin-BOX FD-SOI transistor [7.6]. The thin-BOX FD-SOI MOSFET also allows back-gate control. Since the BOX layer is very thin, V_{th} can be controlled by changing the voltage of the well under the BOX layer. This controllability has the same effect as back-gate bias in bulk CMOS devices. The back-gate bias is applied through a well contact that is formed through the BOX layer. The back-gate of each transistor can be individually controlled, because the well layer is isolated from that of the adjacent transistors by shallow trench isolation (STI). The well is also isolated from the diffusion layer by the BOX layer, so that extra current does not flow from the diffusion layer to the

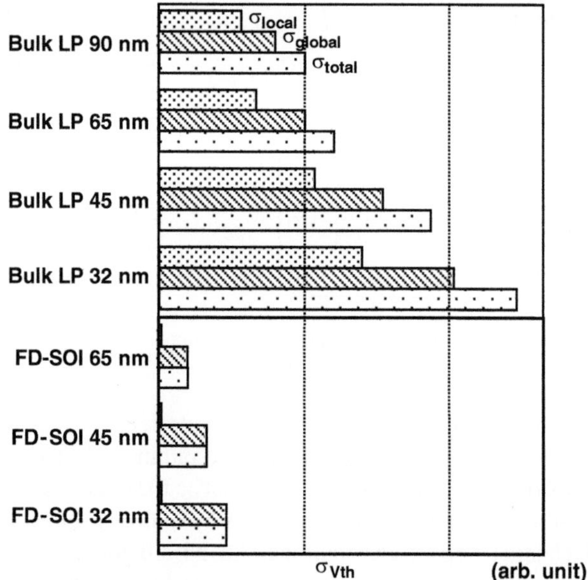

Fig. 7.3 V_{th} variation of conventional bulk transistors and thin-BOX FD-SOI transistors

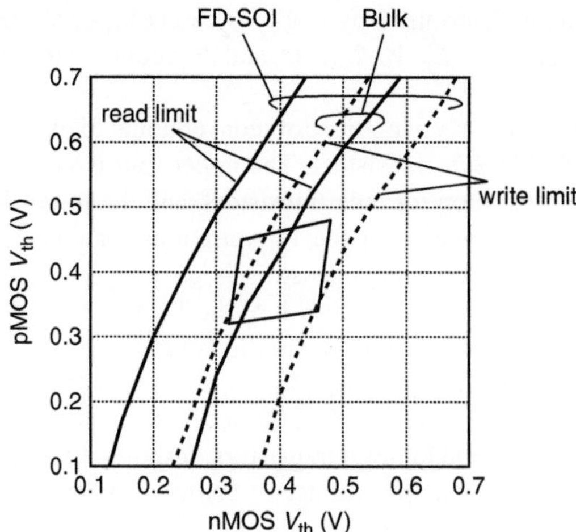

Fig. 7.4 V_{th} window of SRAM cell of 0.7-V V_{DD}

well, even when forward bias is applied. Conventionally, extra leakage current is flown from diffusion layer to substrate (substrate leakage current), and the current prevents the usage of back-gate bias in bulk transistors. This structure solves the problem that various extra currents flow when using back-gate bias in bulk CMOS, and the back-gate bias technique can be easily used in this structure. Figure 7.6 plots

Fig. 7.5 Cross-sectional
SEM micrograph of a
fabricated thin-BOX FD-SOI
transistor [7.6] (© 2008
IEEE)

Fig. 7.6 $V_{gs} - I_{ds}$
characteristics when V_{dd} is
1.0 V and V_{bs} (back-gate to
source voltage) is 0 and 1.0 V

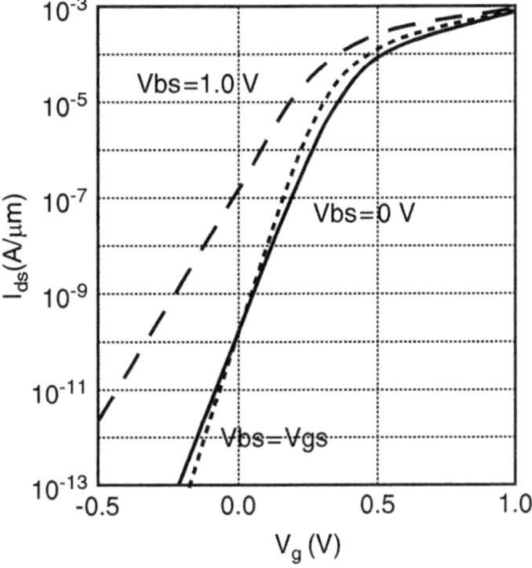

the $V_{gs} - I_{ds}$ characteristics when V_{DD} is 1.0 V and V_{bs} (back-gate to source voltage) is 0 and 1.0 V, which is calculated by device simulation assuming a 65 nm process. Using forward back-gate bias, the I_{ds} (at $V_{gs} = 1.0$ V) can be increased by about 25% without any increase in substrate leakage current, and the I_{ds} (at $V_{gs} = 0$ V) becomes 1,000 times larger. Figure 7.7 plots the measured I_{ds} of a fabricated nMOS device as the back-gate bias is changed. The gate length of this device is 500 nm, and therefore the $V_g - I_{ds}$ characteristics is slightly different from the results of device simulation in Fig. 7.6. However, the controllability of the back-gate bias is clearly proven.

The thin-BOX FD-SOI MOSFETs are applied to SRAM cells. By replacing the bulk MOSFETs with FD-SOI MOSFETs, the lower V_{th} variation realizes high-operating margin for SRAM cell. In addition, the back-gate controllability of proposed MOSFET is effectively used, and the operating margin is improved by changing the memory cell structure. In this section, the margin improved SRAM cell is proposed. Figure 7.8 is the circuit diagram for the proposed SRAM memory cell. The memory cell takes advantage of the thin-BOX FD-SOI structure by adding a feedback mechanism.

The thin-BOX FD-SOI transistor is indicated as a MOSFET with a capacitor which is connected to its back-gate node. The conventional SRAM cell has only

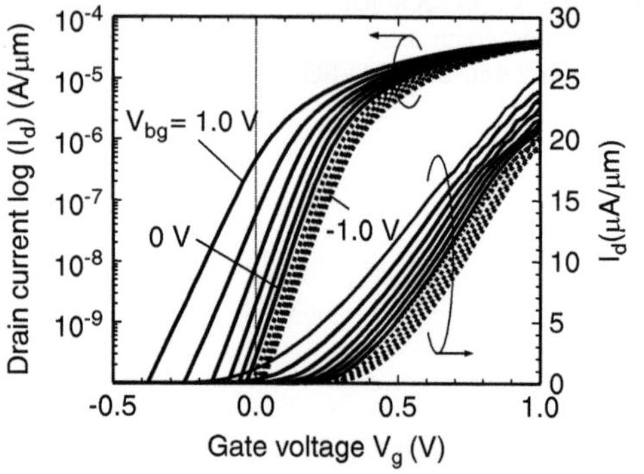

Fig. 7.7 Measured I_{ds} of fabricated nMOS device as back-gate bias is changed

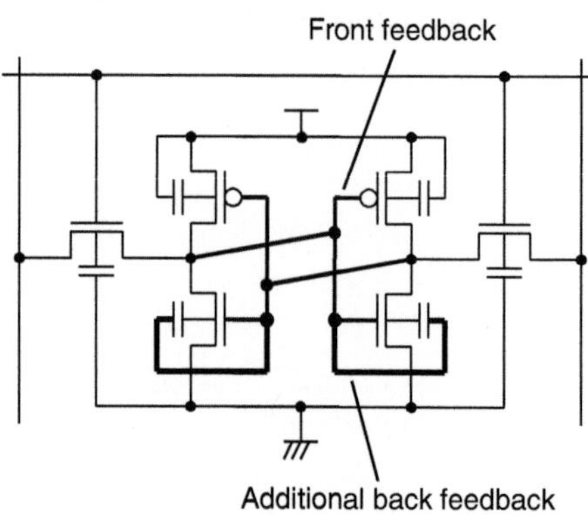

Fig. 7.8 SRAM cell with thin-BOX FD-SOI transistor

front feedback, in which the storage node is connected to the gate node of the load MOS and driver MOS. In the proposed memory cell, a back feedback is added. In the back feedback, the storage node is also connected to the back-gate node of the driver MOS, which is conventionally connected to source node. When the back-gate node is "H" state, the corresponding driver MOS is forward-biased, and the driver MOS acquires a large conductance. This increases the β ratio, and the large β ratio improves read margin. The memory cells can usually retain its data by only using forward feedback, but also using the additional back-gate feedback increases the memory cell stability and compensates for decreased stability due to various factors, e.g. V_{th} variations. As the new feedback is not visible beyond the memory cell, the cell is controlled in the same way as a conventional bulk CMOS memory cell. Furthermore, additional capacitances to control the driver MOS back-gates are attached to the storage node. The capacitance reduces incidence of soft errors. In high-reliability SRAM cells, dedicated capacitors are added to storage nodes with extra process costs. In the proposed SRAM cell, this high reliability is automatically achieved with the new feedback. Figure 7.9 shows butterfly curves for memory cells in which thin-BOX FD-SOI transistors are used with and without back-gate bias. These curves are produced by a hybrid device-and-circuit simulation. Given the device structure, this simulation can analyze physical circuit operation without a

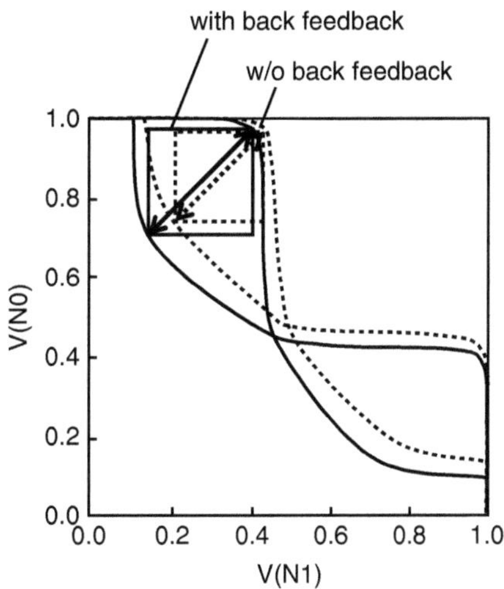

Fig. 7.9 Butterfly curves of SRAM cell with thin-BOX FD-SOI transistors

device model produced by measuring actual devices. The curves show how the use of back-gate bias (the new feedback) improves the SNM. Thus, by applying the thin-BOX FD-SOI to SRAM memory cells, the V_{th} variation is reduced and the operating window becomes wider than that of the window shown in Fig. 7.3 with 20 mV of σ_{vth}. This means that this SRAM cell can operate below a 0.7 V power-supply voltage, because as the back-gate of the driver MOS is controlled and the β ratio increases, the SNM is improved. There is another option when using this thin-BOX FD-SOI transistor. Since read and write operations are achieved by current flowing through the access MOS, a wider access MOS gate leads to higher operating speeds. A wider gate also leads to large access MOS conductance, which enables stable write operation. The conductance of a driver MOS is increased by selectively forward-biasing to back-gate. Therefore, even if the gate width of the access MOS is increased, the SNM can be kept as large as necessary for stable operation. If an RD-cell is used, it might be possible to make the gate width of the access MOS equal to that of the driver MOS without having to increase the memory cell area.

Up to here, an SRAM memory cell composed of six transistors was proposed. Many memory cells are embedded in an SoC, and the memory cell area occupies large area of an SoC. If memory cell area can be reduced, the total SoC area can be greatly reduced. As a method to reduce SRAM cell area, a four-transistor (4-Tr) SRAM memory cell is proposed [7.7]. The cell area is reduced to two-thirds; however, the power consumption of memory cell increases. In this section, a low-power four-transistor memory cell is proposed. Figure 7.10 is a circuit diagram of a conventional 4-Tr SRAM cell, and the figure also shows leakage currents, which are flowing during data retention. AC0 and AC1 work as transfer MOS during cell access and also work as load MOS during data retention. The leakage current for this cell is the total of leakage currents for DR0 and AC1. During data retention, the leakage current of AC0 ($I_{\text{retention}}$) keeps "H" node potential against the leakage

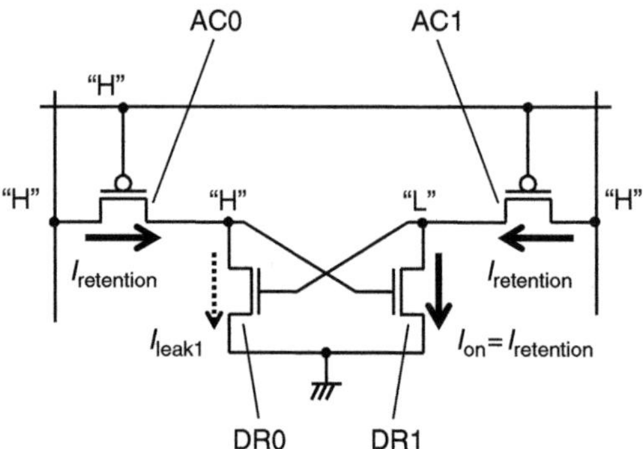

Fig. 7.10 Conventional 4-Tr SRAM cell

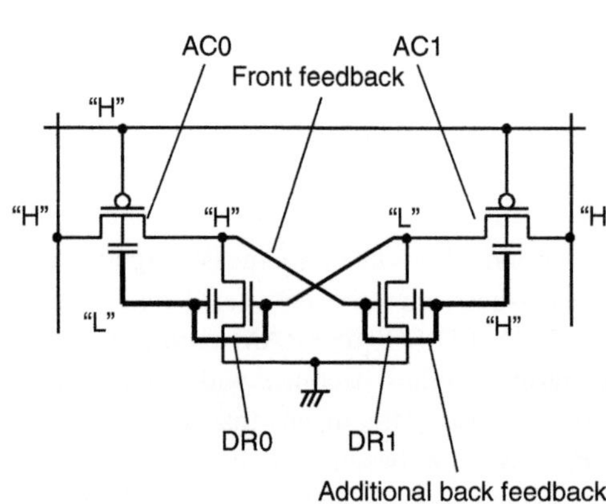

Fig. 7.11 Proposed 4-Tr SRAM cell with thin-BOX FD-SOI transistors

current flowing through DR0 (I_{leak1}). The leakage current of a transistor fluctuates widely because of V_{th} variation. Therefore, the value of $I_{\text{retention}}$ must be designed to be at least $1,000 \times I_{\text{leak1}}$ to keep "H" node potential. At the same time, $I_{\text{retention}}$ flows through AC1. While this current is not necessary for data retention, it still flows and causes a great increase in overall leakage current. The leakage current of a 6-Tr memory cell is at most $3 \times I_{\text{leak1}}$. However, the leakage current of a 4-Tr memory cell is up to $1,000 \times I_{\text{leak1}}$. The cell leakage of a 4-Tr cell is thus about 300 times larger than that of a 6-Tr cell, because unnecessary currents are required for data retention in 4-Tr cells. Figure 7.11 shows a circuit diagram for proposed 4-Tr SRAM cell in which thin-BOX FD-SOI transistors and a new feedback are used. TrMOS1 is forward-biased, and its leakage current $I_{\text{retention}}$ becomes about 1,000 times larger than I_{leak1} since this is required to keep "H" node potential. Since AC1 is not forward-biased, its leakage current I_{leak2} is almost the same as I_{leak1}. Therefore, the cell leakage in this case is no more than $2 \times I_{\text{leak1}}$. Thus, applying a new feedback to 4-Tr cell not only increases cell stability but also dramatically reduces its leakage current. As the feedback is also invisible outside the memory cell like a 6-Tr cell composed of thin-BOX FD-SOI MOSFETs, the peripheral circuits are the same as those for bulk CMOS cells. When thin-BOX FD-SOI MOSFETs are used in these ways described above, the leakage current for a 4-Tr cell is about

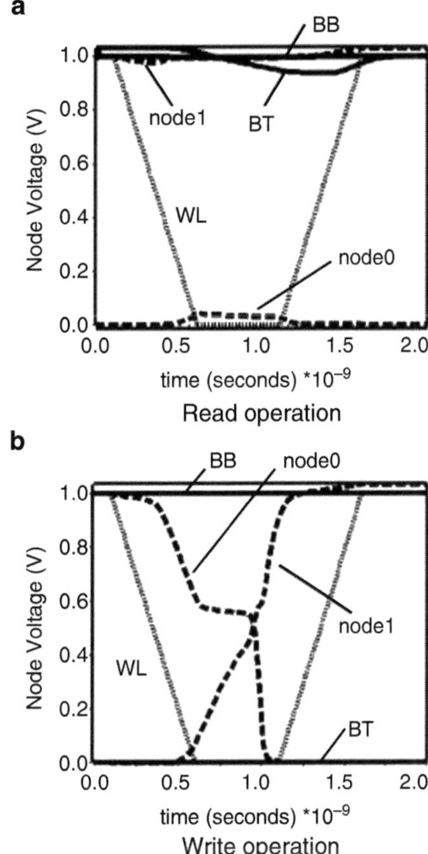

Fig. 7.12 Read and write operation of proposed 4-Tr SRAM cell

two-third of the value for a 6-Tr cell, and is about 0.1% of the value for a 4-Tr cell using bulk transistors. Figure 7.12 shows waveforms of each node voltage of a 4-Tr SRAM cell during read and write operation. The hybrid device-and-circuit simulator is used. The solid waveforms show the bitline voltages, the dotted line shows the wordline voltage, and the broken lines show the storage-node voltage. The read and write operation proceeds correctly. Table 7.1 summarizes the SRAM cell performances with and without the thin-BOX FD-SOI transistors.

7.3   SRAM Cells for FINFET

A FinFET [7.8] is the promising device at the "more-Moore" generation. The FinFET uses 3D structure and is expected to overcome the scaling problems. Figure 7.13 shows the FinFET device. The FinFET is composed of two thin silicon

Table 7.1 Summary of SRAM cell with thin-BOX FD-SOI transistors

	Bulk		FD-SOI	
	6 Tr	4 Tr	6 Tr	4 Tr
Cell area (μm^2)	0.60	0.42	< 0.60	< 0.42
Cell current (A.U.)	1.00	0.50	1.50	0.60
Cell leakage (pA)	55	2200	3.1	2.1
Minimum V_{dd} (V)	1.0	1.0	0.6	0.6
Min.AC power (A.U.)	1.00	0.79	0.36	0.28

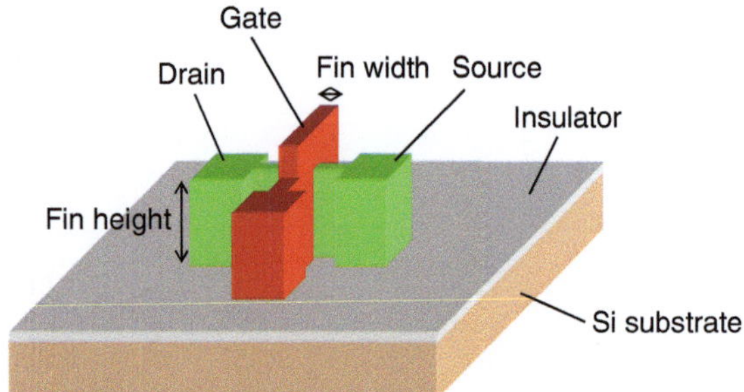

Fig. 7.13 Bird's eye view of FinFET

layers. One layer composes channel layer and its both ends are source and drain electrode. The other layer is gate electrode, which surrounds the channel layer from two sides of channel layer. These layers are formed on an insulator layer and isolated from other transistors. The height of channel layer is called as "fin height" and corresponds to the gate width of conventional bulk transistor. To adjust the fin height of each transistor individually is difficult because the height is decided by manufacturing process. Therefore, the fin height, the gate width of conventional bulk transistor, has to be discrete value using multi-fin structure, which is achieved by forming the channel layer in parallel. The width of the gate layer is called as fin width and corresponds to the gate length of conventional bulk transistor.

The FinFET device has several features. The gate layer controls the channel layer from front and back side of the channel layer; therefore the short channel effect is suppressed and the gate length scaling is easily achieved. This feature also contributes to the reduction of channel dopant. It means that the channel is under the condition of fully depleted. This fully depleted channel contributes to the reduction of channel dopant fluctuation and suppresses the random V_{th} variation, same as the FD-SOI transistor explained in the previous subsection.

Using the FinFET devices, the SRAM cells are also designed. Figure 7.14 shows the SEM micrographs of 6T-SRAM cell using FinFETs [7.9, 7.10]. In the SEM photo, PU, PD, and PG mean load, driver, and access transistors, respectively. These photos show an SRAM with 1.0 β ratio and that with 2.0 β ratio. Conventionally, the β ratio has to be larger than 1.5 to ensure the stable SNM, and the gate width values of access and driver transistors have to be adjusted to suitable value. However, in

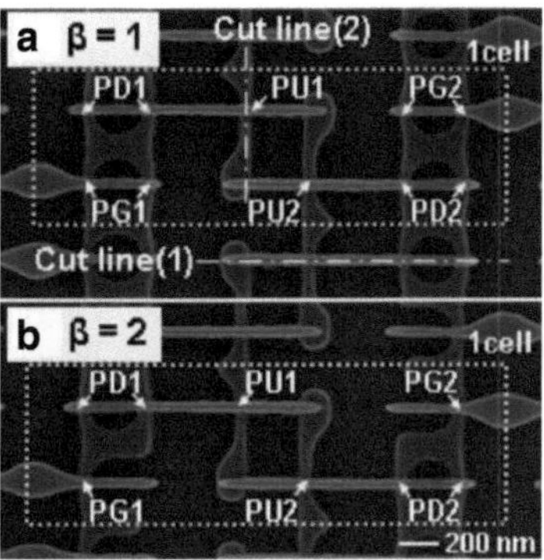

Fig. 7.14 SEM photograph of SRAM cells with FinFET [7.9] (© 2006 IEEE)

FinFET device, the gate width cannot be decided as it should be. Therefore, the 2.0-β ratio cell uses multi-fin structure for access transistor and ensures the β ratio. Furthermore, in FinFET devices, the V_{th} variation is suppressed by fully depleted channel. Therefore, the worst SNM value of FinFET SRAM cell is better than that of conventional bulk transistor cell. This higher worst SNM can eliminate the high β ratio in some cases. The 1.0-β ratio cell can be used in such situation. This FinFET SRAM cell is one of the promising SRAM cells for future manufacturing processes.

References

7.1. K. Takeda, Y. Hagihara, Y. Aimoto, M. Nomura, Y. Nakazawa, T. Ishii, H. Kobatake, A read-static-noise-margin-free SRAM cell for low-VDD and high-speed applications. IEEE J. Solid-State Circuits **41**(1), 113–121 (2006)

7.2. Y. Morita, H. Fujiwara, H. Noguchi, Y. Iguchi, K. Nii, H. Kawaguchi, M. Yoshimoto, An area-conscious low-voltage-oriented 8T-SRAM design under DVS environment. in *IEEE VLSI Circuits Symposium Digest*. June 2007, pp. 256–257

7.3. H. Noguchi, Y. Iguchi, H. Fujiwara, Y. Morita, K. Nii, H. Kawaguchi, M. Yoshimoto, A 10T non-precharge two-port SRAM for 74% power reduction in video processing. in *Proceedings of the IEEE Computer Society Annual Symposium VLSI (ISVLSI)*. March 2007, pp. 107–112

7.4. B. Cheng, S. Roy, A. Asenov, The scalability of 8T-SRAM cells under the influence of intrinsic parameter fluctuations. in *Proceedings of the IEEE European Solid-State Circuits Conference (ESSCIRC)*. September 2007, pp. 93–96

7.5. Y. Morita, H. Fujiwara, H. Noguchi, Y. Iguchi, K. Nii, H. Kawaguchi, M. Yoshimoto, Area comparison between 6T and 8T SRAM cells in dual-Vdd scheme and DVS scheme. IEICE Trans. Fundam. **E90-A**(12), 2695–2702 (2007)

7.6. Y. Morita, R. Tsuchiya, T. Ishigaki, N. Sugii, T. Iwamatsu, T. Ipposhi, H. Oda, Y. Inoue, K. Torii, S. Kimura, Smallest Vth variability achieved by intrinsic silicon on thin BOX (SOTB) CMOS with single metal gate. in *IEEE VLSI Technology Symposium 2008, Digest of Technical Papers*. June 2008, pp. 166–167

7.7. K. Noda, K. Matsui, K. Imai, K. Inoue, K. Tokashiki, H. Kawamoto, K. Yoshida, K. Takeda, N. Nakamura, T. Kimura, H. Toyoshima, Y. Koishikawa, S. Maruyama, T. Saitoh, T. Tanigawa, A 1.9 − μm2 loadless CMOS four-transistor SRAM cell in a 0.18 − μm logic technology. in *IEEE IEDM Technical Digest*. December 1998, pp. 643–646

7.8. D. Hisamoto, W.-C. Lee, J. Kedzierski, H. Takeuchi, K. Asano, C. Kuo, E. Anderson, T.-J. King, J. Bokor, C. Hu, FinFET–A self-aligned double-gate MOSFET scalable to 20 nm. IEEE Trans. Electron Devices **47**, 2320–2325 (2000)

7.9. A. Nackaerts, M. Ercken, S. Demuynck, A. Lauwers, C. Baerts, H. Bender, W. Boulaert, N. Collaert, B. Degroote, C. Delvaux, J.F. de Marneffe, A. Dixit, K. De Meyer, E. Hendrickx, N. Heylen, P. Jaenen, D. Laidler, S. Locorotondo, M. Maenhoudt, M. Moelants, I. Pollentier, K. Ronse, R. Rooyackers, J. Van Aelst, G. Vandenberghe, W. Vandervorst, T. Vandeweyer, S. Vanhaelemeersch, M. Van Hove, J. Van Olmen, S. Verhaegen, J. Versluijs, C. Vrancken, V. Wiaux, M. Jurczak, S. Biesemans, A 0.314μm2 6T-SRAM cell build with tall triple-gate devices for 45nm node applications using 0.75NA 193nm lithography. IEDM Tech. Dig. 269–272 (2004)

7.10. H. Kawasaki, K. Okano, A. Kaneko, A. Yagishita, T. Izumida, T. Kanemura, K. Kasai, T. Ishida, T. Sasaki, Y. Takeyama, N. Aoki, N. Ohtsuka, K. Suguro, K. Eguchi, Y. Tsunashima, S. Inaba, K. Ishimaru, H. Ishiuchi, Embedded bulk FinFET SRAM cell technology with planar FET peripheral circuit for hp32 nm node and beyond. in *VLSI Technology, Digest of Technical Papers*. 2006 Symposium on 2006, pp. 70–71

Index

Access conflict, 76
Access MOS, 54, 56, 57, 133
Access MOSFETs, 5
Access-NMOS, 76, 84
Access time, 9, 31
Access transistor, 7, 62
Aggressive shrinkage, 80
Alpha particle, 90, 94, 95, 97, 111
Alpha particle soft-error, 94, 95
Alpha ray, 90, 97, 103
Alpha ray soft-error, 94
Angle parameter, 21
Application processor, 32
Area overhead, 118
Array boost technique, 44, 54–59
Array structure, 5
Ary-, 71
Ary-VDM, 67, 69–72
Assist circuits, 43, 72
Asymmetric, 62
Asymmetry, 9
Asynchronous, 30
Asynchronous SRAM, 30
ATD pulse, 30
Auger recombination, 112

Back-body bias, 29
Back-gate, 131–133
Back-gate bias, 131–133
Back-gate control, 129
Back-gate feedback, 132
Bandgap, 28
Battery effect, 90, 105, 111, 112
Bipolar, 90, 101, 105, 107–112, 116
Bit lines, 5, 7, 26
Bitline shifting circuitry (BSC), 77

BL shifter, 85
β ratio, 9, 51, 57, 76, 81, 132, 133, 136, 137
β-ratio RD-cell, 57
Butterfly curb, 9
Butterfly curves, 13, 39, 54, 59, 61, 76, 132

Cache, 43, 51–53
Cache data array, 46
Cache memory, 10
CADDETH, 109
Capacitive WAC, 67
CD shift, 64, 65
Cell current fluctuation, 48
Cell stability, 57, 76, 127
Channel dopant, 129
Charge collection, 90, 99, 100
Charge deposition density, 113
Charging current, 14
CMOS, 95, 115, 118
CMOS 6-transistor SRAM memory cell, 11
Coefficients, 21
Column control circuits, 5
Column multiplexer, 7
Column-select line, 8
Common, 76
Conflicting relationship, 15
CORIMS, 93, 99–102
Critical charge, 94, 98, 102, 103
Critical dimension (CD) shift, 64
Crosstalk, 39
CYRIC, 119

Data array, 51
Data storage node, 13
Data upset, 9

K. Ishibashi and K. Osada (eds.), *Low Power and Reliable SRAM Memory Cell and Array Design*, Springer Series in Advanced Microelectronics 31, DOI 10.1007/978-3-642-19568-6, © Springer-Verlag Berlin Heidelberg 2011

DAVINCI, 112, 113
Design for manufacturability (DFM), 80, 81
Deviation, 21
Device scaling, 90
Diagonal line, 9
Die-to-die V_{th} variation, 16
Different, 75
Diffusion layer, 38
Dmy-VDM, 67, 69, 71, 72
Dopant distribution, 16
DP-SRAM cell, 44, 75–77, 79–85, 87
DP 8T cell, 84
DRAM, 25, 91, 94, 95, 102, 103
Drive-NMOS, 76, 79, 81, 84
Driver MOS, 54, 56, 57, 132, 133
Driver MOS back-gates, 132
Driver MOSFET, 5
Driver transistor, 82, 136
Drive transistor, 87
Dual-port, 75
Dual-port SRAM, 44, 74, 75
Dummy bitline, 47
Dummy cells (DCs), 43, 47, 48
Dummy-cell current fluctuation, 48
Dummy-column cell, 43
Dummy-column cell (DC) and edge-column
        cell (EC), 47
Dummy word line, 47, 48

ECC, 90, 91, 93, 94, 100, 105, 115–117, 119
Edge-column cell, 43
Electrical β ratio, 54, 56, 57, 76, 81
Electrical charge, 14
Electrical imbalance, 38
Electrical stability, 44
Electric field, 28
Electric field relaxation (EFR) scheme, 29
Electric field strength, 28
Electric stability, 43
Embedded, 32
Experimental FBC, 22

Fail bit count, 22
Failure, 90
Failure rate, 22
FD-SOI, 136
FD-SOI MOSFETs, 131
FinFET, 125, 135–137
First metals, 38
Flip-flop, 5
Flipped, 13
Flow chart, 22

Fluence, 93
Four-dimensional spherical expressions, 21
Four-transistor (4-Tr) SRAM memory cell, 133
Full CMOS 6-transistor, 5
Fully depleted, 136, 137
Funneling, 98, 100, 111, 113
Funneling length, 95

Gate controller (GC), 64
Gate-induced drain leakage (GIDL) currents,
        26, 84
Gate leakage, 84
Gate-oxide tunnel leakage, 25, 26
Gate-tunnel leakage currents, 26
GIDL currents, 26
Global, 83
Global variations, 16, 59, 83
Gradient, 21

Half-selected, 76
Half select issue, 128
Hamming code, 117
High-R cell, 1
High stability, 33

Impact ionization, 112
INC, 100
Infinitesimal deviation, 21
Inverter, 5

Junction tunnel leakage, 26

LANSCE, 102
Leakage current, 26
Line edge roughness (LER), 16
Lithographically symmetrical (LS) cell, 37
Load MOS, 132, 133
Load MOSFET, 5
Load-PMOS, 74
Local and global components, 16
Local bus, 30
Local variation, 43, 83
Local V_{th} variation, 59–62, 83, 125, 127, 128
Logical threshold voltage, 13
Lombardi's surface mobility model, 112
Lower-voltage-range operation, 36
Lowest aspect ratio, 38
Low leakage, 33
Low-voltage operation, 37

MBU, 91, 94, 100, 115
MBU MCUs, 93
MBU, SEFI, SEL, 92
MCBI, 90, 91, 105, 112, 113, 115
MCU, 91, 93, 94, 100, 101, 105–108, 112, 113,
     115
MCU/MBU, 100
Memory array, 5
Memory cell design, 18
Memory cell V_{SS} line (V_{SSM}), 30
Microprocessors, 36
Minimum operating voltage (VDD$_{min}$), 20,
     125
Minimum supply voltage, 22
MNT, 92, 115
Mobile phones, 25
Monoenergetic, 93
Monoenergetic neutron, 101
Monte Carlo simulation (MC), 19
Moore's law, 2, 125
More-Moore, 135
MOSFET, 96, 98, 100
MOSFET V_{th} variation, 16
Multi-fin structure, 137

Negative, 52
Negative bias, 29
Negative V_{BB} back-body bias, 28
Negative V_{GS}, 28
Neutron fluence, 119
Neutrons, 90, 93, 97–100, 103, 105, 108, 115,
     119
130-nm manufacturing process, 33
90 nm node, 40

Operating margins, 15
Optical proximity correction (OPC), 38

Parasitic capacitance, 39
Parity bits, 117, 118
Passive resistance elements, 63
Pattern fluctuation, 38
PCSE, 92, 115
Phase shift, 38
Photolithography misalignment, 38
Point of symmetry, 38, 40
Poly-silicon layer, 38, 40
Portable electronic devices, 36
Positive, 52
Precharge circuit, 5
Precharged, 30
Precharged bit line, 13

Predecoder, 50
Process corners, 16
Process scaling, 16
Process variation, 34
Pull-down NMOSs (drive-NMOS), 74, 127
Pull-up, 127
Pull-up current, 14
Pull-up PMOSs, 74, 128
PVT, 64
p-well, 108, 111, 113, 116

Quasi-monoenergetic, 115
Quasi-monoenergetic neutron, 99

RAC, 64, 65, 71, 77, 85
Random dopant fluctuation (RDF), 16
RAT, 62, 64–66
RD-cell, 57, 58
Read, 7
Read and WACs, 74
Read and write limit lines, 17
Read assist circuit (RAC), 62
Read failure, 13
Read margin, 59, 60, 125
Read operation, 7, 9, 13
Read (SNM) FBC, 24
Read stability, 76, 81, 127, 128
Read stability, SNM, 66, 76
Rectangular diffusion (RD) cell, 57
Refresh, 9
Reliability, 59
Replica access transistors, RATs, 62, 66
Resistive division, 13
Retained data, 13
Retention margin, 71
Retention mode, 29
Row address comparator (RAC), 76
1RW, 74, 80
1R1W, 75
2RW, 75, 81
2RW dual-port, 80
2RW dual-port SRAM, 75

SBU, 91, 97, 106, 115
SEB, 92
SECIS, 93, 99
Secondary ions, 94, 108, 113
Second metals, 39
SEE, 90, 91, 93
SEFI, 92, 101
SEGR, 92

SEL, 92
Selectively, 133
Sense amplifier, 7, 46
Sensitivity analysis, 20
SER, 95, 97, 99, 101, 105, 115, 119
SER–SECIS, 99
SESB, 92, 113
SET, 92
SEU, 90, 92, 93, 98, 102, 103
SEU cross section, 93, 99
Short channel effect, 16
Single-ended, 75
Single-ended bitline, 128
Single-port, 74
Single-port SRAM, 74
Size of 6T cell, 2
Skew, 46
Snake pattern, 81
SNM, 13, 43, 54–57, 59–62, 76, 77, 81, 133,
        136, 137
SNM-limit line, 17
SoC, 18
Soft-errors, 90, 94–98, 103, 109, 110, 113,
        115, 132
Source bias circuit, 43
Source line voltage control techniques, 29, 33
Spallation, 90, 94, 97–100, 103, 113
Spallation neutron, 93
SPICE parameter, 22
SRAM capacitance, 22
SRAM cell, 5
SRAM DC margin, 20
SRAM operating margin analyses, 16
SRAM operation, 11
SRAM retention stability, 33
SRH, 112
Stability, 43, 125
Stability assisting circuits, 43
Stable DC function, 22
Standard deviation, 16
Standby leakage, 84
Standby power dissipation, 25, 29
Static noise margin (SNM), 9, 46, 76, 125
Static-noise margin: read margin (SNM), 17
STI, 98, 100, 112
Subthreshold leakage currents, 26, 28
$3\text{-}\sigma V_{\text{th}}$, 55
Symmetrical, 54
Synchronous DP-SRAM, 44
System-on-chip, 5

7T, 125
Tag array, 51

Taylor expansion, 21
6T cell, 2
8T-cells, 75
8T DP-cell, 44, 75, 76, 80, 81, 84
8T DP-SRAM, 81
Temperature, 59
Terrestrial neutron, 90, 97, 99, 104, 113
Thin, 129
Thin-BOX FD-SOI, 125, 128, 129,
        131–135
Threshold voltages, 26, 45
Transfer MOS, 133
Transfer NMOSs (access-NMOS), 74
Transfer PMOS, 127
4-Tr cells, 134, 135
6-Tr cells, 134, 135
8T read margin-free cell, 125
10T read margin-free cells, 125
4-Tr memory cell, 134
6-Tr memory cell, 134
4-Tr SRAM cell, 133–135
6T single-port SRAM cells, 125
TSL, 105
6T SP-cell, 80, 81, 84
6T SP-SRAM, 81
6T SRAM, 80, 81
8T SRAM, 75, 80, 81
10T SRAM, 128
6T SRAM bitcell, 128
6T SRAM cell, 72, 81, 125, 127, 128, 136
7T SRAM cell, 125
8T SRAM cell, 81, 127, 128
10T SRAM cell, 127, 128
8T SRAM (DP) cell, 84
6T SRAM (SP) cell, 84
Typical V_{th} value, 21

UHD DP-SRAM, 85
UHD-8T-SRAM cell, 81, 82
Ultra-high-density (UHD) DP-SRAM, 84

VDD$_{\text{min}}$, 74, 125
Voltage-adapted pulse, 50
Voltage-adapted timing-generation scheme, 43,
        46
Voltage generator, 54
V_{th} combinations, 19
V_{th} curve, 70, 83
V_{th} distribution of manufactured devices, 18
V_{th} operating window, 17
V_{th} variations, 16, 62, 128, 129, 131–134, 136,
        137

V_{th} window analysis, 16
V_{WL}, 62–66, 71

WAC, 70, 71
Well taps, 52
Wide and thin, 81
Wide-range-voltage operation, 46
WL driver, 62
WL level, 61, 62
WL voltage, 63, 66
WL voltage level, 62
Word decoder, 5
Word driver, 50
Word line, 5, 26

Word line lowering, 43
Worst-case model, 19
Worst-case SNM and write operation analysis, 17
Worst vector, 22
Worst V_{th} combination, 20
Write amplifier, 7, 8
Write assist circuit (WAC), 67
Write failure, 13
Write limit line, 17
Write margin, 60, 67, 70, 71, 83
Write operation, 7, 8
Write stability, 76
Write-trip-point, 70
Writing margin, 17